MW01099056

The Learning Equation®

INTERMEDIATE ALGEBRA

Student Workbook

THOMSON

™

BROOKS/COLE

Australia • Canada • Denmark • Japan • Mexico • New Zealand • Philippines
Puerto Rico • Singapore • South Africa • Spain • United Kingdom • United States

Printed in Canada
1   2   3   4   5   6   7   07   06   05   04   03

Printer: Webcom

ISBN: 0-534-42034-6

For more information about our products, contact us at:
Thomson Learning Academic Resource Center
1-800-423-0563

For permission to use material from this text, contact us by:
Phone: 1-800-730-2214
Fax: 1-800-730-2215
Web: http://www.thomsonrights.com

Brooks/Cole—Thomson Learning
10 Davis Drive
Belmont, CA 94002-3098
USA

Asia
Thomson Learning
5 Shenton Way #01-01
UIC Building
Singapore 068808

Australia/New Zealand
Thomson Learning
102 Dodds Street
Southbank, Victoria 3006
Australia

Canada
Nelson
1120 Birchmount Road
Toronto, Ontario M1K 5G4
Canada

Europe/Middle East/South Africa
Thomson Learning
High Holborn House
50/51 Bedford Row
London WC1R 4LR
United Kingdom

Latin America
Thomson Learning
Seneca, 53
Colonia Polanco
11560 Mexico D.F.
Mexico

Spain/Portugal
Paraninfo
Calle/Magallanes, 25
28015 Madrid, Spain

The Learning Equation® 3.5

Brooks/Cole Publishing
Thomson Learning

2004 All rights reserved.

Bob Pirtle
Publisher, Mathematics

Jennifer Huber
Senior Editor, Developmental Mathematics

Patrick J. Call
Chief Technology Officer

Julia Pluss
TLE Project Manager

Christopher Delgado
Technology Project Manager

Kirsten Markson
Development Editor

Content Review Panel

Alan Tussy
Professor of Mathematics
Citrus College
Glendora, California

Mark D. Clark
Associate Professor of Mathematics
Palomar College
San Marcos, California

John E. Daggett
Math Lab Coordinator
DeAnza College
Cupertino, California

Beth Hempleman
Math Learning Center Coordinator
Mira Costa College
Mira Costa, California

Tim Hempleman
Adjunct Faculty
Palomar College and California State University
San Marcos, California

Harris S. Shultz
Professor of Mathematics
California State University
Fullerton, California

Melvin L. Hamburger
Lecturer, Mathematics
Laramie County Community College
Cheyenne, Wyoming

Sally Jackman
Professor of Mathematics
Richland College
Dallas, Texas

William John Radulovich
Professor of Mathematics
Florida Community College
Jacksonville, Florida

Julane B. Crabtree
Professor of Mathematics
Johnson County Community College
Overland Park, Kansas

Sylvia Carr
Lecturer, Mathematics
Southwest Missouri State University
Springfield, Missouri

Janet E. Teeguarden
Associate Professor
Ivy Technical State College
Indianapolis, Indiana

Roseanne B. Benn
Associate Professor, Mathematics
Prince George's Community College
Largo, Maryland

iii

Multimedia Development

ACERRA

Executive Director
Kirsten Robinson

Senior Main Engine Programmer
Reed Carriere

Development Programmer
Geoffrey Heaton

Project Manager
Kevin Kernohan

Development Programmer
Collin Chenier

Testing Manager
Diana Guy

Senior Creative Design
Mike Martel

Authoring

Bill Currie	*Collin Chenier*	*Geoffrey Heaton*	*Jo-anne Landriault*
Kevin Kernohan	*Lianne Zitzelsberger*	*Madelyn Hambly*	*Mike Cherun*
	Catherine Fitzpatrick	*Mike Martel*	*Duane Kennedy*

Audio Production
Bartmart Audio

Narration Voices

June Dewetering	*Patrick Fry*	*Ron Purvis*	*Genevieve Spicer*
Lisa Thompson	*Russ Faubert*	*Kevin Kernohan*	

Lunar Logic Inc.
BCA Integration
Michael Jones

Editorial Development

First Folio Resource Group, Inc.

Project Manager
David Hamilton

Senior Editor
Eileen Jung

Storyboard Editors

Jackie Lacoursiere	*Darren McDonald*	*Don Rowsell*

Picture Research and Copy Editing

Robyn Craig	*Susan Marshall*

Text Design, Layout and Art

Tom Dart	*Claire Milne*	*Quinn Banting*

Math Explorers

Design and Programming
Ron Blond

Windows Versions
Grant Arnold
Vladislav Hala

Product Development

Garry Popowich	*Larry Markowski*	*Ted Keating*
Ron Blond	*Eleanor Milne*	*Chris Kirkpatrick*
Milt Petruc	*Maxine Stinka*	*Ralph Lee*
Sharon Tappe	*Joanne Adomeit*	*Bob Robinson*
Dale Weimer	*Steven Daniels*	*Brenda Stewart*
Lisa Wright	*Lee Kubica*	*Frances I. Tallon*
Angus McBeath	*Barb Morrison*	*Stella Tossell*
Tom Winkelmans	*Connie Goodwin*	*Susan Woollam*
Paula Allison	*Theresa Gross*	*Alan Tussy*
Duncan McCaig	*Terri Hammond*	*Henry Decell*
Katharine D. Tetlock	*Peggy Hill*	*Matthew O'Malley*
Barry Mitschke	*Carol Besteck Hope*	

Authors

Karen Auch	*Chris Kirkpatrick*	*Kevin Rowan*
Leo Boissy	*Jean MacEachern*	*Rod Rysen*
Mark Bredin	*Melani McCasin*	*Selina Samji*
Lana Chow	*Greg McInulty*	*Rick Sept*
Gloria Conzon	*Roxane Menssa*	*Joe Shenher*
Brent Corrigan	*Lee Mitchell*	*Brenda Stewart*
Colin Garnham	*Keith Molyneux*	*Katharine D. Tetlock*
Mike Gmell	*Loretta Morhart*	*Colleen Tong*
Connie Goodwin	*Rob Muscoby*	*Terry Wallace*
Barry Gruntman	*Kanwal I.S. Neel*	*Dale Weimer*
Christine Henzel	*David Nutbean*	*Ketri Wilkes*
Joe Hirschegger	*Robert Payne*	*Alan Tussy*
Carol Besteck Hope	*Brent Pfeil*	*Henry Decell*
Terry Imhoff	*Grant L. Plett*	*Matthew O'Malley*
Ted Keating	*Caleb Reppenhagen*	

Contents

Lessons

PREFACE

The Learning Equation® *(TLE)* is interactive multimedia courseware and student workbooks for developmental mathematics, from basic math/arithmetic through intermediate algebra. Developed in Macromedia Director®, TLE is attractive with a professional look and feel, and runs fast and reliably on most Windows and Macintosh platforms.

TLE uses interactive, activity-directed learning to involve the student in their own education. Each TLE lesson is designed to build skills in algebra and problem solving. The entire series of lessons has a sound, curriculum-based foundation. As students progress through the lessons, they will learn the vocabulary of mathematics, practice key concepts, and develop their skills in reasoning, modeling, and analysis.

Each TLE lesson contains a wealth of application problems. Numerous applications are included from disciplines such as business, entertainment, science and technology, and history.

TLE Student Guide

Introduction

The Learning Equation (TLE) is a multimedia courseware and workbook package developed as an online learning tool for developmental mathematics. If your instructor has adopted TLE, you will log in through a program called BCA. BCA is a course management program that will allow you to access information on your course anytime online, do the lessons, and turn in assigments online.

TLE is designed to engage you in your own learning while building skills in algebra and problem solving. As you progress through the lessons, you will practice key concepts, build your vocabulary, and develop your basic math and reasoning skills.

TLE PIN Codes

You will need two PIN codes to access TLE. The PIN code that comes packaged with your TLE Student Workbook gives you access to the TLE Courseware, including the Lessons, Practice, Self Checks, and Resources.

In order to access information specific to your course, such as your assignments and syllabus, you will need to ask your instructor for a course PIN. Your instructor will be able to give you a course PIN once a course in BCA has been set up.

Logging In

To login to TLE, you will need to go through BCA because the TLE content is housed within the BCA system. TLE refers to the interactive learning content that you will work through, while BCA is the system that will keep track of your assignments and grades, as well as allow you access to course specific information from your professor.

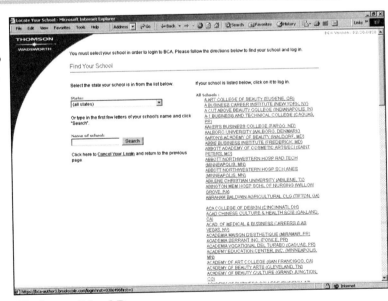

Find your School Page

Initial Login

❶ Open your Web browser and type **http://BCA.brookscole.com** into the address field.

❷ Click on **First Time User**.

❸ Make sure that the name of your school appears in the "School" field. If it does not, click on "Find Your School" and follow the next steps.

❹ The "Find Your School" screen will appear.
In the "State" field, select your state from the drop-down menu.

❺ In the "Name of School" field, type your school's name and click on **Search**. A list of schools will appear at the right.

❻ Click on your school name in the list.

❼ In the "PIN Code" field of the "First Time Users" page, type the instructor's PIN code for your course and/or section. Click on **Register**.

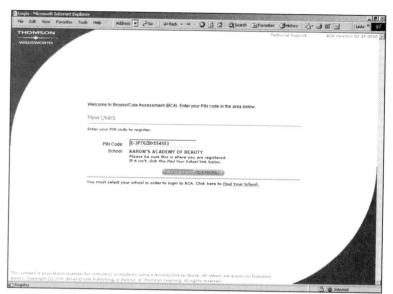

First Time Users Page

Completing the Registration Form and Creating a Login Name (or User Name) and Password

❶ On the registration page ("Welcome to BCA Registration"), type the appropriate information in all fields. Fields marked with a red asterisk must be filled in.

Note: The program automatically creates a login name (or user name) for you. You may change this login name by deleting the one that is supplied and typing a new one. Please make sure your login name and password are easy to remember. Record your account information in a safe place. Avoid using spaces in your login name, and remember that login names and passwords are case-sensitive. (You'll have to type them the same way, including any capitalization, each time.) In the future, you will need to use your login name and password, not the PIN code, to log in.

❷ After the information has been entered, click on **Register and Begin Brooks/Cole Assessment**.

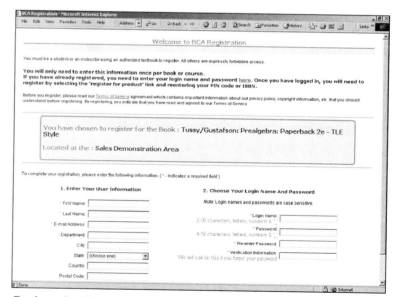

Registration Page

Subsequent Login

❶ Go to **http://BCA.brookscole.com**.

❷ Click on **Login**. (If you already logged in to the BCA site on your computer, your school name should automatically appear on the login ("Registered Users") screen. If it does not appear, click on Find Your School and follow steps 3–5.

❸ Type your user name (or login name) and password; then click on **Login**.

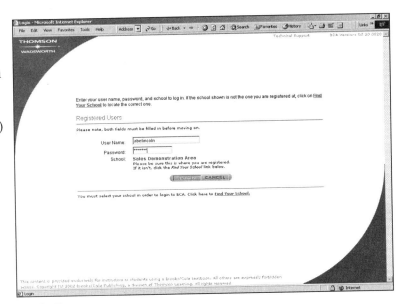

Login Page

Getting Started

You are now ready to get started with BCA. Note the menu that runs along the left side of the screen. This menu, which is always evident, provides access to all the course areas, including "My Assignments," "Progress," and "Tutorials," as well as opportunities to change your password or register for an additional course or product.

In order to access your TLE assignments, you will need to register for the TLE content. To do this, click on **Enter Content Access Pin** in the left navigation menu and enter the PIN code you received with your book. Click **Submit**. You are now ready to access your TLE assignments.

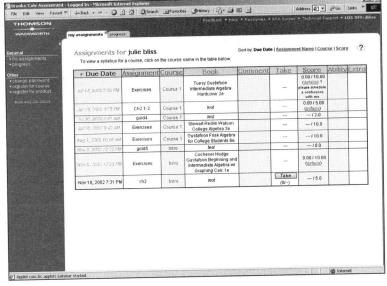

My Assignments Page

Logging Out

To log out, go back to the Home page. Click the LOG OUT link in the upper right of the page. When the BCA Front Porch screen appears, you know you have successfully logged off of your TLE session.

It is important to remember to log out to end your TLE session. Failure to log out will record incorrect time-spent information and could result in a poor grade for the work you have done.

Log Out Link Page

Moving Around in TLE

Lesson Title

The objectives and prerequisites for the lesson and basic navigation instructions are presented in the on-screen tabbed notebook. Click on individual heading tabs.

Select the NEXT button to start the **Introduction**.

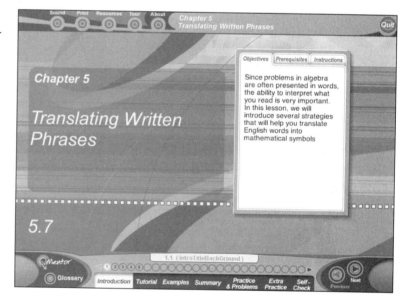

Menu Bar

The MENU bar has five drop-down options.

The sound can be adjusted to individual needs. Use the drop-down **Sound** menu from the menu bar to turn the sound off, rewind the narration, or to set the level from 1 to 7. Headsets may also be used with the sound feature.

The **Print** button will send an image of the current screen to your printer.

Under **Resources**, you'll find links to additional programs called **Explorers**, calculators, and other tools. In order to use these, you will need to download them from the sidebar menu on the BCA Start Up screen.

The **Tour** button will show you how to use the TLE program and the **About** button will tell you about the creators of TLE.

You can also access the Glossary at any time by selecting the **Glossary** button on the lower portion of the screen. Within each lesson, the Glossary is also available by clicking words highlighted in blue ("hot text"). Once in the Glossary, select the word of your choice from the alphabetic menu. Return to the lesson by selecting the **Close Window** button or use the **Alt-F4** keys.

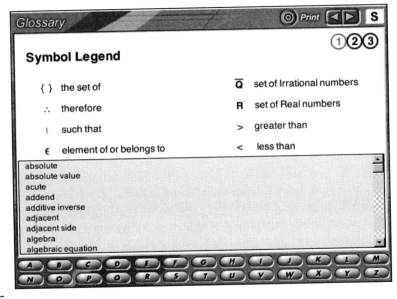

Moving Around a Lesson

Navigation Bar

You always know where you are in a lesson by looking at the top TLE Navigation Bar. The chapter number and lesson title can be found at the top of the screen, and the lesson information at the bottom of the screen.

Chapter 1
Data in Tables 2

The TLE Navigation Bar is designed to be easy to use, with seven icons to reference the lesson components.

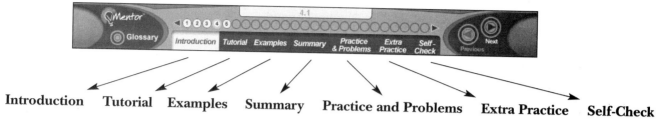

Introduction   Tutorial   Examples   Summary   Practice and Problems   Extra Practice   Self-Check

Click the appropriate icon to go to the lesson component of your choice.

Pages

The number of "pages" in each lesson component is also displayed in the Navigation Bar. Several screens of material may be included in a "page," which represents a "complete thought."

A yellow border highlights the current page of the lesson component. Once a page is completed, it is highlighted in red.

A user can jump to individual pages by clicking the page number on the Navigation Bar.

Previous and Next

Navigate through the lesson by selecting the NEXT button.

The PREVIOUS button allows you to go back to review earlier screens.

TLE Components

The Learning Equation® Intermediate Algebra consists of 64 lessons that cover the college intermediate curriculum. Lessons are designed to take about 90 minutes but you can go at your own pace.

Every lesson consists of the following seven components.

Introduction

The opening screen briefly outlines the lesson and its objectives. Prerequisites for the lesson and instructions are available from an on-screen tabbed notebook.

To start the lesson, select the NEXT button on the opening screen.

The **Introduction** opens the lesson with a problem set in the context of a career, a real world or consumer experience, or a game. You will use knowledge you already have to solve the proposed problem. The **Introduction** is short, motivating, and highly interactive.

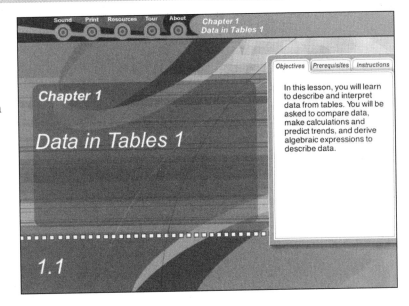

Tutorial

The TLE **Tutorial** offers the main instruction for the topic of the lesson. This section is brief and straightforward, with simple examples that helps you understand the concepts. The **Tutorial** is intended to encourage to learn by doing.

By providing different solutions or different routes through the same problem, TLE is designed to help you realize that there can be many ways to solve a problem.

TLE uses **Hints** to remind you when you can use strategies you've already learned. **Success Tips** offer ways to build confidence or to support independent study and learning.

Feedback Boxes
You can move any feedback box by selecting the green bar at the top of the box, and dragging it to the desired location. You can hide, and then reveal, the content of the feedback box by double clicking on the green bar. To close the feedback box, select the button in the upper left corner of the green bar.

Examples

TLE **Examples** expand on what you learned in the **Tutorial**. The **Examples** are highly interactive and include feedback to responses, hints, or alternative solutions.

Up to 12 Examples may be presented in each lesson. A **hidden picture** is progressively revealed as you complete each example. Once the picture has been completely revealed, you may select a **Picture Information** button to learn more about it.

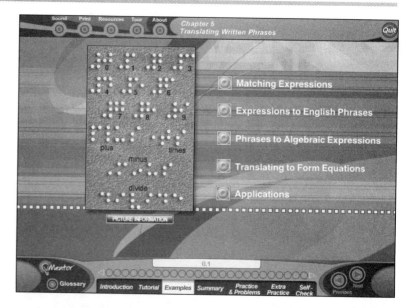

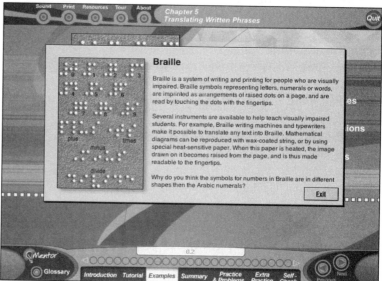

Summary

The TLE **Summary** revisits the **Introduction** and encourages you to apply the mathematics learned in the **Tutorial** and **Examples** to the problem in the **Introduction**.

Practice and Problems

The TLE **Practice and Problems** presents 20 or 25 questions. They are organized in four or five categories with five questions in each category. You can click on any box to try a question.

As in the **Examples**, another **hidden picture** is revealed when questions are completed correctly.

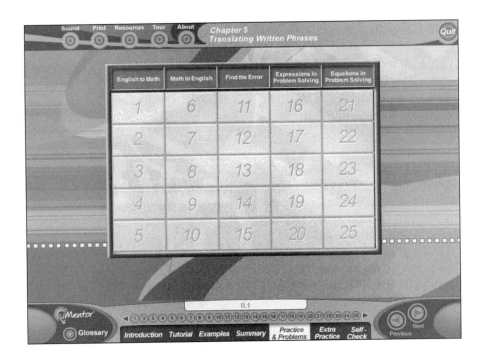

Extra Practice

The TLE **Extra Practice** presents questions like the ones in the **Examples.** After each question, you have the option of seeing a sample solution, retrying the question, or seeing the correct answer if they were incorrect.

Questions for each type of exercise are dynamically generated from a **mini-data bank** of up to 60 questions.

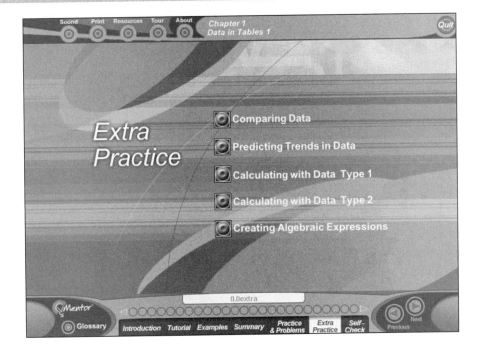

Self-Check

The TLE **Self-Check** presents up to **10 questions** like those in the **Extra Practice**. There are **three unique Self-Checks** for each lesson.

After seeing your scores, you can **review the questions** and their **answers** one at a time, **try again**, or **see** the **correct answer** or a **sample solution**.

On achieving a minimum standard (usually about 70%), the lesson is considered completed.

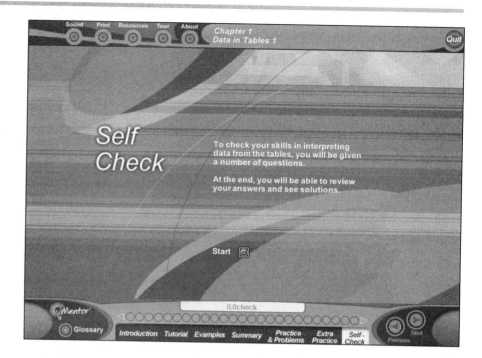

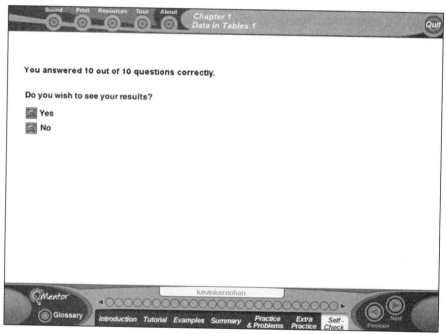

Feedback Boxes

You can move any feedback box by selecting the blue bar at the top of the box, and dragging it to the desired location. To close the feedback box, select the yellow button in the upper left corner of the the blue bar.

Success Tips for Users

You learn by doing. The following suggestions will facilitate this process:

1. **Take notes:** Write down an example or two that was worked out in the **Tutorial** to keep as a model for other problems throughout the lesson. Also write down any steps found in the **Summary** or rules provided in the **Tutorial**.

2. **Use the Glossary:** Key terms are highlighted in blue throughout the lesson. Find the meaning of these terms in the Glossary. To access the Glossary, either click on the term itself or select the **Glossary** button on the lower portion of the screen. Keep a list of the highlighted terms, along with an example. Such a list is helpful while working through a lesson and also aids in review at the end of a lesson.

3. **Discuss:** Whenever possible, discuss what you have learned or what you do not understand with a partner.

4. **Review the Examples:** The **Examples** section usually provides different information from that covered in the **Tutorial**. Review each category in the **Examples**.

5. **Use calculators and other tools:** Though you can complete some examples and problems using mental math skills, you should also, where appropriate, use other tools such as pencil and paper, diagrams, and calculators, to solve out a problem.

6. **Be aware of levels of difficulty in Practice and Problems:** Recognize that the categories of questions in the **Practice and Problems** tend to increase in difficulty from left to right across the columns, and from top to bottom within each column or category.

7. **Further enrichment:** Follow up on the Picture Information provided for each completed Hidden Screen in the **Examples** and in the **Practice and Problems** components.

8. **Use pencil and paper:** When completing the **Extra Practice** and the **Self Check** components, take time to transfer information from screen to paper. When answers are reviewed, your work can be compared with sample answers for completeness and accuracy.

9. **Use the Self-Check:** The **Self-Check** can be used in different ways:

   (a) as a placement tool – it can be attempted before the start of a lesson to determine whether the lesson material is new to you.

(b) to check for understanding – it can be attempted at the
end of a lesson to check that you have understood the
lesson material.

A poor mark on the **Self-Check** indicates that you should be
directed back to the **Tutorial** and the **Examples** components of
the lesson, or alternatively, to the **Extra Practice** section.

Achievement of at least 70% on **Self-Check** produces a check mark
indicator in the box to the left of the lesson on the MENU.

10. **Create an instruction manual:** Create your own guide on how to
use different functions and features of this program.

11. **Screen dumps:** On Macintosh computers, it is possible to take
a "screen dump" (a picture of what is on the screen) by pressing
the keys **Shift-Command-3** simultaneously. In Windows, press
Alt-Prnt Scn.

When using the Explorers, you may wish to take screen dumps for
later reference as well as write about any observations you made
about features not mentioned in the software.

TLE Features

Interacting with TLE Courseware

You can interact with TLE in a variety of ways.

Interaction 1

You may input **simple text** and press RETURN after each input.

Every effort has been made to anticipate all reasonable forms in which you might enter answers. Text answers usually consist of numbers, and sometimes variables. Other alphabetic entries are rarely required, in order to avoid entering mis-spelled but mathematically correct answers.

Different numerical forms of answers may be allowed, such as 1/2, 0.5, 0.50, and so on. However, if all the information in a problem is expressed in one form, like whole numbers, the answers are usually expected to be in the same form.

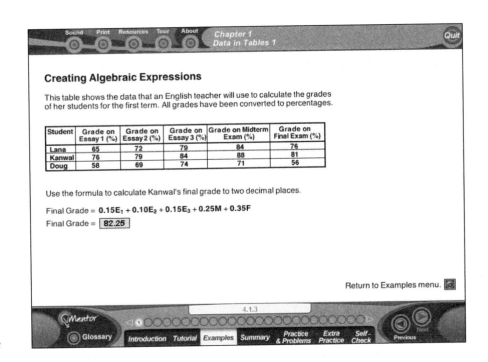

Interaction 2

You may click a button for a multiple choice response.

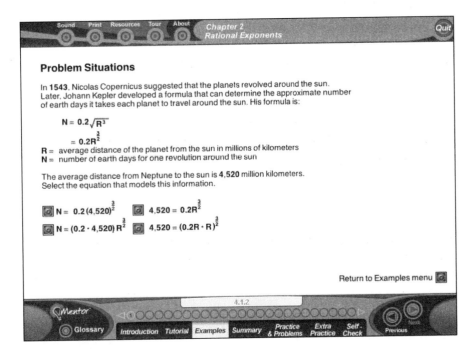

In some multiple choice questions, you may select more than one response, and then press a DONE button. You can then reflect on your choice(s).

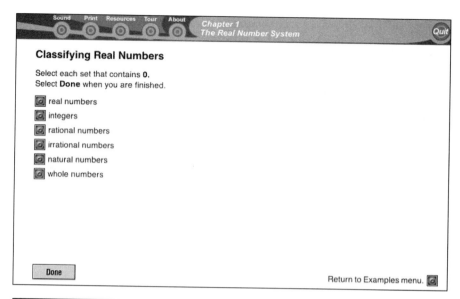

Interaction 3

You may drag one or more items to the appropriate location(s).

Click on the item and drag it with the mouse. Correctly placed items stick; incorrectly placed items bounce back.

You should drag and place the items carefully with the mouse. If you are careless about the position of an item in a box, a correct item may bounce back and appear to be incorrect.

For some "drag-and-drop" questions, you may drag more than one item to the appropriate location and then press a DONE button.

After any of the above interactions, select NEXT to continue.

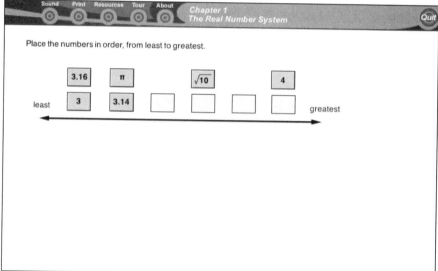

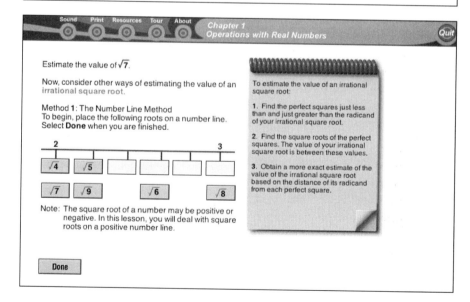

Sound

A soundtrack accompanies parts of TLE with a **sound control (Off or 1 to 7)**, which is available to you. Narration accompanies the Introduction, Tutorial, and Summary components.

Explorers

In several TLE lessons, you can use supplementary software called **Explorers**. They are designed to **encourage** you **to explore** mathematics concepts, skills, and procedures, **test** your ideas, and **reflect** on your actions.

The Explorers can be found under Resources. In order to use these, you will need to download them from the sidebar menu on the BCA Start Up screen.

To install the TLE Explorers, follow these instructions:

On the sidebar menu from the BCA Start Up screen, click **Manage TLE Data**.
On the page that opens, click the link to **Install TLE Explorers**. Follow the installation instructions. Launch TLE Online and the resources will be available.

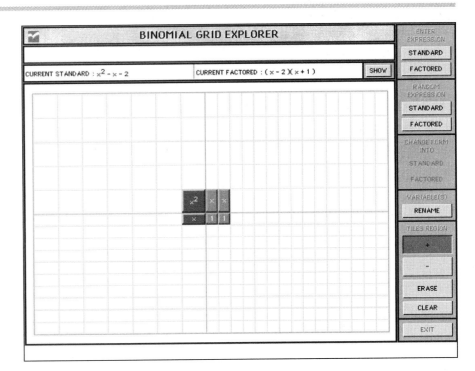

Glossary

TLE allows you to **access** an **on-screen** Glossary using **hot text** or a pull-down **menu**. Hot text is high-lighted on screen and can be clicked for direct access to the Glossary.

The Glossary provides complete definitions and examples. It helps you understand **the language of mathematics** and is just one of the many features that promote math connections within mathematics and to other disciplines.

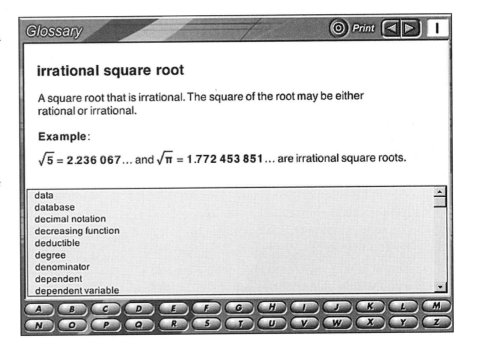

While the course content is contained in TLE, course management and access to syllabi and assignments is housed in BCA. Below are instructions on how to use key areas in the BCA program to keep track of your assignments, grades, and progress in the course.

The "My Assignments" Area in BCA

The "My Assignments" page provides a chart (or syllabus) that lists the assignments and allows access to them. In this chart, you can see, for any assignment:

- Due date and time
- Course name
- Comments from the instructor (if any)
- **Take** button and number of takes allowed and completed
- Assignment name
- Book (test or text for the tutorial)
- Scores
- Extra credit (if any)

Check the chart regularly to keep up with the due dates and times of assignments.

The "Take" column includes a **Take** button to get started on each assignment (see "Access Assigned Homework" below). The "Take" column also identifies how many attempts were made and how many attempts are allowed on each assignement. For instance, "(0/–)" indicates that you have not attempted the assignment and that you may attempt it in an unlimited number of times. Similarly, "(0/1)" indicates that you have not attempted the assignment and that you may attempt it only once.

Access Assigned Homework

If your instructor has assigned tutorials or homework problems, you will see those listed in the chart on the "My Assignments" page. Click on the **Take** button in the chart to get started on an assignment. Any work completed in the assigned tutorial is graded and recorded in the instructor's gradebook.

Access Assigned Quzzes and Tests

If your instructor has assigned quizzes or tests, then you will see those listed in the chart on the "My Assignments" page.

❶ To get started, click on the assignment's **Take** button. *Note:* Do not click on **Take** until you are ready to attempt the assignment. Clicking on **Take** but not finishing a quiz or test can lower your score.

❷ On the screen that appears, click on **Click Here**. The assignment will open.

❸ The screen for your assignment is organized as follows:

- **Previous and Next Tabs.** By clicking on these tabs at the top of the page, you can navigate forward and backward between questions.

- **Jump to Question.** By choosing from the drop-down menu at the top of the page, you can jump from one question to another or click on **End Test** to finish.

- **Timer.** If your instructor has set a time limit for the assignment, the timer will be active and it will indicate the amount of time remaining for the assignment.

- **Question text.** Be sure to read the question carefully.
- **Submit Button or Answer Field.** Depending on the type of question, the method of entering answers will differ.

The "Result Details" Chart

The "Result Details" chart will appear after you have completed an assignment. The chart is a record of your assignment and includes the following important information:

- Assignment name, overall score, times taken, and time spent.
- Item, score, possible score, and right/wrong answers for each question.

Of greatest importance are details available in the "Right/Wrong" column. You should use this material to reinforce what you have done correctly and to learn from your mistakes.

To review your results:

❶ In the "Right/Wrong" column of the chart, click on **Right** or **Wrong** next to any item to view the original question, your answer, and the correct answer.

❷ At the top right, click on **Return to Assignment List**. The "My Assignments" page will open, and you can see how your work has been recorded.
  a. In the "Take" column, your attempt is now recorded;
    that is, "(0/–)" has become "(1/–)."
  b. In the "Scores" column, your score is now recorded, and the details link appears.

You can click on the **details** link to access the "Result Details" chart at any time.

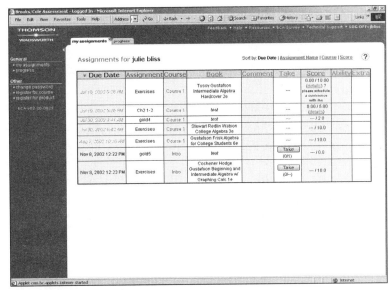

An Assignment Screen

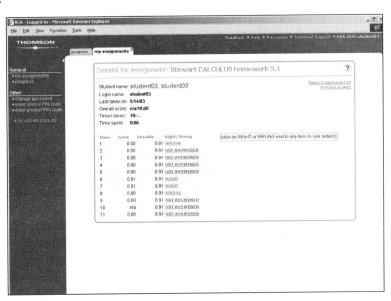

Result Details Chart

The "Progress" Area

The "Progress" page lets you check on your personal gradebook.

❶ From the "Start Up" page, click on **Progress** under the "General" option in the menu at the left of the screen. A chart will open. The chart serves as a personal gradebook and contains the following:

- Course
- Assignment name
- Assigned date
- Due date and time
- Taken-on date
- Possible score
- Extra credit (if any)
- Score
- Notes from the instructor (if any)

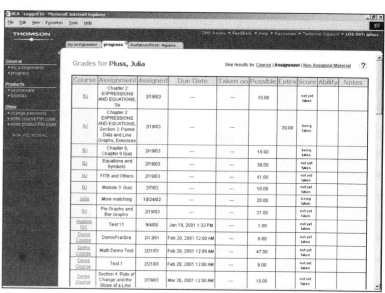

Gradebook on the "Progress" Page

❷ The chart can be printed. Beneath the chart, click on **Display in separate window for printing, saving, or email**.

❸ In the browser's menu, click on **File**; then click on **Print**.

Changing Your Password

To change your password, click on **Change Password** under "Other" in the menu at the left of the screen. Change your account details, and then click on **Change It**.

Registering for Additional Courses and Products

Situation: You have already created an account, and now an instructor for another course gives you a new PIN code.

❶ Click on **Enter Course Code** under "Other" in the menu at the left side of the screen.

❷ In the "PIN Code" field, enter the PIN code your instructor has given you; then click on **Submit**.

❸ You will see the "Congratulations" screen. You will see assignments for this new course in "My Assignments."

Situation: You have already created an account, and now you have a new textbook that also uses BCA Tutorials.

❶ Click on **Enter Content Access PIN Code** under "Other" in the menu at the left side of the screen.

❷ In the "PIN Code" field, enter the PIN code packaged with your book or the ISBN number on the back cover of your new book, then click on **Submit**.

Other Resources

When you need help, use the links in the upper right corner:

- **FeedBack.** Use this form to let us know how BCA is working for you.
- **Help.** Get context-sensitive help that makes sure you have the answers you need.
- **Resources.** Access glossaries and tools for your discipline.
- **Technical Support.** Go to the BCA Technical Support Website for FAQs, student guides, information on new products, downloads, Report a Bug and feedback forms, JRE (Java Runtime Environment) instructions, and contact information for Online Help.

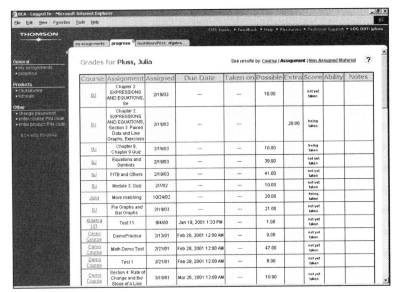

BCA Technical Support Website

Online Tutoring with vMentor

Access to BCA also means access to online tutors and support through vMentor, which provides live homework help and tutorials. To access vMentor while you are working in TLE, click on the **vMentor Tutoring** button in the bottom left corner above the glossary.

Next, click on the **vMentor** button; you will be taken to a Web page that lists the steps for entering a vMentor classroom. If you are a first-time user of vMentor, you might need to download Java software before entering the class for the first class. You can either take an Orientation Session or log in to a vClass from the links at the bottom of the opening screen.

All vMentor Tutoring is done through a vClass, an Internet-based virtual classroom that features two-way audio, a shared whiteboard, chat, messaging, and experienced tutors.

You can access vMentor Sunday through Thursday, as follows:

> 5 p.m. to 9 p.m. Pacific Time
>
> 6 p.m. to 10 p.m. Mountain Time
>
> 7 p.m. to 11 p.m. Central Time
>
> 8 p.m. to midnight Eastern Time

If you need additional help using vMentor, you can access the Participant Guide at this Website: **http://www.elluminate.com/support/guide.pdf.**

The Learning Equation®

INTERMEDIATE ALGEBRA

Student Workbook

REAL NUMBERS

1.1 Data in Tables I

Organizing numerical data in a table can help you identify trends, patterns, and relationships. When you express patterns and relationships algebraically, you can use the expressions to fill in and extend the data.

Example 1

The table shows the number of wins, losses, ties, and points for teams in the National Hockey League Western Conference at one point during a recent season.

Team	W	L	T	Points
Detroit	35	9	4	74
Colorado	26	14	9	61
Chicago	25	15	11	61
Toronto	22	19	9	53
St. Louis	21	20	8	50
Phoenix	21	24	4	46
Vancouver	17	20	12	46
Los Angeles	17	22	11	45
Calgary	18	23	9	45
Edmonton	18	25	6	42
Anaheim	17	27	5	39
Dallas	14	24	10	38
San Jose	11	35	4	26

What percent of its total games played did the first-place team win? the last-place team?

Solution

The teams are arranged in the table according to the number of points. Detroit is in first place, with 74 points. San Jose is in last place, with 26 points.

Detroit had 35 wins, 9 losses, and 4 ties, so the team must have played 35 + 9 + 4 or 48 games in all.

San Jose had 11 wins, 35 losses, and 4 ties, so the team must have played 50 games in all.

To find the number of wins as a percent of total games played, convert the ratio $\frac{wins}{total\ games}$ to a percent.

For Detroit:

$\frac{35}{48} = 35 \div 48 = 0.729$ or about 73 hundredths or 73%.

For San Jose:

$\frac{11}{50} = 11 \div 50 = 0.22$ or 22 hundredths or 22%.

Detroit won about 73% of their games and San Jose won about 22% of theirs.

Example 2

Use the table from Example 1. Points are calculated based on the number of wins and ties. How many points is a win worth? a tie?

Write an algebraic expression to describe how to calculate a team's total points.

Solution

Each team's points must be calculated the same way. Since San Jose has the least points, it may be the easiest one to work with.

Use trial and error to find the pattern.

First try: If wins and ties were worth 1 point each, San Jose would have 11 points for wins + 4 points for ties, or 15 points in all. Since San Jose has 26 points, either wins or ties or both must be worth more than 1 point.

Second try: If wins and ties were worth 2 points each, San Jose would have 22 points for wins + 8 points for ties, or 30 points in all. 30 is closer to 26, but this is still not correct.

Third try: Since wins are harder to get than ties, they may be worth more points. If wins were worth 2 points and ties were worth 1 point, San Jose would have 22 points for wins + 4 points for ties, or 26 points in all. This seems to be the correct pattern.

Check the pattern by using statistics for another team.

Detroit has 35 wins and 4 ties, so it should have 70 + 4 or 74 points. It does, so the pattern seems to be correct.

Write the algebraic equation.

A team scores 2 points for each win (w) and 1 point for each tie (t).

$$\text{Total points} = 2w + 1t$$
$$= 2w + t$$

Test the expression with a different team.

For Dallas:

$$\text{Total points} = 2w + t$$
$$= 2(14) + 10$$
$$= 38$$

This equation can be used to calculate total points for any NHL team.

Exercises

1. Define a trend. Give an example from the lesson.

2. Describe a type of numerical data that could be organized in a table. Sketch a table to show how you would arrange the data.

3. The table shows prices for identical items at three different food stores.

	Price Way	Snack Mart	Food Plus
milk ($/qt)	1.29	1.00	1.38
bread ($/loaf)	2.49	1.59	1.49
ground beef ($/lb)	2.19	2.39	2.99
tuna ($/can)	0.99	0.95	1.04
tomato sauce ($/8 oz can)	0.39	0.29	0.25
coffee ($/8 oz jar)	3.59	4.55	3.99
dog food ($/4 lb bag)	3.59	3.99	4.09
laundry soap ($/92 oz box)	7.89	8.68	7.55

(a) Give some examples of trends you can see in the data.

(b) Where would you shop if you wanted to buy only milk? if you wanted to buy every item on the list?

(c) Describe an algebraic formula you could use to help you calculate each store's price for an order that includes 3 lb of ground beef, 2 cans of tomato sauce, and 4 loaves of bread.

(d) Which store offers the best price for the order described in part (c)?

4. The table shows changes in regional hog prices from 1960 to 1982.

Year	Price per hundredweight ($)
1960	18.40
1965	22.80
1970	26.00
1975	39.30
1978	56.50
1979	69.40
1980	62.00
1981	56.30
1982	56.10

(a) What trends can you find in the price data?

(b) Why does the price change from $56.50 to $69.40 represent a more significant increase than the price change from $39.30 to $56.50?

(c) Which year had the highest selling price?

(d) In which year was the price equal to about 57% of the highest selling price? about 37%?

(e) What price would you have predicted for 1983? Why?

5. The table shows the number of hours worked by a group of software designers.

Name	Mon	Tues	Wed	Thurs	Fri
Brooks	9.0	7.5	6.5	9.5	8.5
Clark	7.0	7.0	7.0	7.0	7.0
Jao	8.5	8.0	10.0	9.5	9.0
Nichols	10.0	11.0	9.5	10.5	10.0
Pierce	9.0	9.0	10.0	9.0	8.0
Tyler	8.5	10.5	12.0	8.0	8.0

   (a) Add a column to the table to show the total number of hours each person worked.

   (b) Calculate the average number of hours worked this week.

   (c) Suppose that an employee is paid overtime for every hour worked after the first 40, and that employees are paid 1.5 times as much for overtime hours as for regular hours. If w dollars represents the regular hourly wage of an employee, create an algebraic expression the company can use to calculate an employee's gross pay, for any employee who works overtime.

6. Use the hockey table in Example 1. How would the order of the teams change if wins were worth 3 points and ties were worth 1 point?

7. Create a problem of each type based on the data in the hockey table in Example 1. If you wish, you can use data from a current sports almanac to update the statistics.

   (a) comparing data

   (b) calculating with data

   (c) looking for patterns

1.2 Data in Tables II

In this lesson, you worked with **recursive tables**, where rows of data rely on information from previous rows. You interpreted the data shown in tables, described data using words and algebraic expressions, and made calculations and predictions.

Example

A fishing camp borrows money to buy boat motors. The owner plans to make one regular payment at the end of each summer, with the option of making an extra payment if the season is good.

Year	Opening Balance ($)	Interest Charged ($)	Regular Payment ($)	Extra Payment ($)	Closing Balance ($)
1	86,000.00	6,020.00	20,974.60	–	71,045.40
2	71,045.40	4,973.18	20,974.60	–	55,043.98
3	55,043.98	3,853.08	20,974.60	–	37,922.46
4	37,922.46	2,654.57	20,974.60	–	19,602.43
5	19,602.43	1,372.17	20,974.60	–	0.00

The table outlines the loan repayment plan:

Write an analysis of the data that addresses each question:
(a) Does the table show recursive or nonrecursive data? Explain.
(b) What is the period of the loan?
(c) How much did the fishing camp borrow?
(d) How much is still owing at the end of the third year?
(e) What is the interest rate of the loan?
(f) How much of Year 3's payment goes to pay interest and how much to repay the principal?
(g) Suppose that at the beginning of Year 5 the interest rate on the fishing camp loan had increased to 8.5%. How much would the camp still owe at the end of Year 5, if the owner made only the same regular payment?

Solution

(a) The table shows recursive data. Each opening balance depends on the previous year's closing balance.

(b) The loan will be completely repaid at the end of 5 years, so the period of the loan is 5 years.

(c) The opening balance in the first year, $86,000, represents the amount borrowed, or the principal.

(d) At the end of Year 3, the table shows that $37,922.46 is still owing on the loan.

(e) To find the interest rate, look at the interest charged in the first year.
$$\frac{\text{interest charged}}{\text{opening balance}} = \frac{6,020}{86,000} = 0.07 = \frac{7}{100} = 7\%$$

(f) The payment in Year 3 is $20,974.60. Of that amount, $3,853.08 represents the interest charged for that year. The rest of the payment is used to pay down the principal.
$20,974.60 - 3,853.08 = \$17,121.52$
The amount directed to the principal is $17,121.52. This represents the difference between the opening balance for Year 3, and the opening balance for Year 4.

(g) The table shows the changes that would result from the increased interest rate.

Year	Opening Balance ($)	Interest Charged ($)	Regular Payment ($)	Extra Payment ($)	Closing Balance ($)
5	19,602.43	1,666.21	20,974.60	–	294.04

Exercises

1. Explain how you can tell whether a table shows recursive or nonrecursive data.

2. Define each term.
   (a) principal

   (b) interest rate

   (c) loan period

3. Explain how to determine the interest rate on a loan if you know the principal and the amount of interest charged in one year.

4. Winnie wants to buy a leather jacket priced at $279.95. A clerk explains that a sale will begin on Friday and continue for five days. On each day of the sale, the price of the merchandise will be reduced by 10%.
   The table shows how the jacket price will change during the first four days of the sale:

Day	Price ($)	Tax 1 ($)	Tax 2 ($)	Total ($)
Thursday	279.95	19.60	14.00	313.54
Friday	251.96	17.64	12.60	282.19
Saturday	226.76	15.87	11.34	253.97
Sunday	204.08	14.29	10.20	228.57
Monday	183.68	12.86	9.18	205.72

   (a) Winnie cannot afford to spend more than $250, but if she waits too long, someone else may buy the jacket. Identify the earliest day when Winnie can afford to buy the jacket.

   (b) What is the Tax 2 rate in Winnie's state? Explain how you know.

   (c) Write an expression in words that describes Friday's price as a percent of Thursday's price. Use pre-tax prices.

   (d) Write an algebraic expression that describes Friday's price as a percent of Thursday's price. Let P_1 represent Thursday's price and P_2 Friday's price.

(e) The sale continues until Tuesday. If no one buys the jacket, what will the total price be by then?

(f) Winnie decides to wait until Tuesday to buy the jacket. When she arrives, a sign announces, "Save even more! We pay Tax 2 today!" How much does she end up paying for the jacket?

5. The Reid family has saved $3,300 for a trip to the Grand Canyon. They plan to invest their savings, adding an additional $400 per month for the next year. The table outlines the first six months of their saving plan.

Month	Opening Balance ($)	Deposit ($)	Interest Earned ($)	Month-end Balance ($)
Jan	3,300.00	–	16.50	3,316.50
Feb	3,316.50	400.00	18.58	3,735.08
Mar	3,735.08	400.00	20.68	4,155.76
April	4,155.76	400.00	22.78	4,578.54
May	4,578.54	400.00	24.89	5,003.43
June	5,003.43	400.00	27.02	5,430.45

(a) What is the January interest as a percent of the initial investment? This represents the interest rate.

(b) How much interest will the investment earn from January until the end of March?

(c) Write an algebraic expression you can use to calculate the amount of interest the account will earn in any month. Let O represent the opening balance.

(d) What is the opening balance for July?

6. Use the table from the example.

   Determine the extra payment at the end of Year 2 that would pay off the loan at the end of Year 4. (Assume the regular payment amount stays the same, and no extra payment is made in Years 3 and 4.)

   Hint: How can you use the closing balance in Year 4 to find the opening balance in Year 4? the opening balance in Year 3?

7. The table provides data on a student loan. No payments will be made while the student is at school for the next three years.

Year	Opening Balance ($)	Interest Rate (%)	Interest Charged ($)	Year-end Balance ($)
1	5,200.00	9.00	468.00	5,668.00
2	5,668.00	8.50	481.78	6,149.78
3	6,149.78	8.00	491.98	6,641.76

   (a) Describe the recursive relationship in the table.

   (b) Write four questions about the table, similar to questions posed in the lesson. Record the answers to your questions, then exchange questions with a classmate.

1.3 The Real Numbers

In this lesson, you learned to **classify numbers** according to the systems to which they belong:
- **Natural** numbers are the counting numbers 1, 2, 3, ….
- **Whole** numbers include the set of natural numbers and 0.
- **Integers** include the negative counting numbers as well as all whole numbers.
- **Rational** numbers can be expressed as the quotient of one integer divided by another, where the denominator is not equal to 0. In decimal form, these numbers either terminate or repeat.
- **Irrational** numbers cannot be expressed in fraction form. In decimal form, they are nonterminating and nonrepeating.
- **Real** numbers include all rational and irrational numbers.

Example

Which letter on the diagram represents the smallest possible number system that contains each set?

$\{-1, 3, -5, 7, -9, ...\}$ $\{100, 200, 400, 800, 1{,}600, ...\}$

$\{0, 2, 4, 6, 8, 10, ...\}$ $\left\{\dfrac{1}{9}, \dfrac{2}{9}, \dfrac{3}{9}, \dfrac{4}{9}, ...\right\}$

$\{0.3, 0.09, 0.0081, ...\}$ $\left\{\sqrt{2}, \sqrt{3}, \sqrt{5}, \sqrt{7}, \sqrt{11}, ...\right\}$

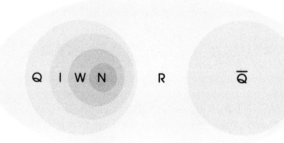

Solution

$\{-1, 3, -5, 7, -9, ...\}$ is composed of both positive and negative counting numbers, or integers. The letter I represents integers.

$\{100, 200, 400, 800, 1{,}600, ...\}$ is composed of positive counting numbers starting at 100. The letter N represents natural numbers.

$\{0, 2, 4, 6, 8, 10, ...\}$ is composed of counting numbers starting from 0. The letter W represents whole numbers.

$\left\{\dfrac{1}{9}, \dfrac{2}{9}, \dfrac{3}{9}, \dfrac{4}{9}, ...\right\}$ is composed of fractions, where the numerator and denominator are non-zero integers. The letter Q represents rational numbers.

$\{0.3, 0.09, 0.0081, ...\}$ is composed of decimal fractions, where each decimal terminates. Terminating decimals are rational. The letter Q represents rational numbers.

$\left\{\sqrt{2}, \sqrt{3}, \sqrt{5}, \sqrt{7}, \sqrt{11}, ...\right\}$ is composed of square roots of prime numbers. In decimal form, these roots neither terminate nor repeat. Nonterminating, nonrepeating decimals are irrational. The symbol $\overline{Q}$ represents irrational numbers.

Exercises

1. Which number systems contain the set {5, 10, 15, 20, 25, ...}?

2. Identify any three irrational numbers.

3. Show where you would locate each number along a number line.
  (a) −2.75

  (b) $-\dfrac{27}{8}$

  (c) $-\pi$

  (d) $-\sqrt{14}$

  (e) −3.1212...

  (f) −3.989889888...

  (g) −4.1296496...

  (h) −2.11141114...

4. State the smallest number system that has each number from Problem 3 as a member.

5. Match each expression to its graph.

   (i) $-\sqrt{5} \leq x \leq \sqrt{11}$, x is an irrational number

   (ii) $x \geq 6$, x is a natural number

   (iii) $\frac{1}{2} \leq x \leq \sqrt{24}$, x is a real number

   (iv) $0 \leq x \leq 4$, x is a whole number

   (v) $-1 < x \leq 2$, x is a rational number

   (vi) $-2 \leq x \leq 1$, x is an integer

(a)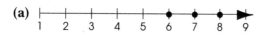

(b)

(c)

(d)

(e)

(f)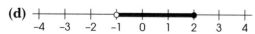

Note: A solid dot indicates that a number is part of the set. An open dot indicates that a number limits a set, but is not part of it. A solid bar indicates that an infinite number of points are included along its length.

6. Calculate the decimal equivalent for each square root where this is possible.

   (a) $\sqrt{49}$ **(b)** $\sqrt{361}$ **(c)** $-\sqrt{121}$

   (d) $\sqrt{7}$ **(e)** $\sqrt{42}$ **(f)** $\sqrt{-72}$

7. Is each number rational or irrational? Give a reason for your answer.

   (a) 324   **(b)** $-\sqrt{5}$ **(c)** -2π

   (d) $\sqrt{81}$ **(e)** $\frac{3}{16}$ **(f)** 15.125

   (g) 43.84821...   **(h)** $-12.333...$ **(i)** 4.050050005...

8. List the square roots of the natural numbers from 1 to 10.

   (a) Which numbers have square roots that are natural numbers?

   (b) Which numbers have irrational square roots?
   Hint: None of the decimal numbers would show repeating patterns if you could see more digits.

9. List the square roots of the rational numbers from 0.01 to 0.10.

   (a) Which numbers have rational square roots?

   (b) Which numbers have irrational square roots?

10. List the square roots of the rational numbers from 0.1 to 1.0.

   (a) Which numbers have rational square roots?

   (b) Which numbers have irrational square roots?

   (c) Why is this pattern different from the one you found in Problems 8 and 9?

11. Here is a method you can use to locate irrational numbers on a number line.
   (a) Draw a number line from 0 to 8 on grid paper.
   (b) Draw a square along the line so one side of the square is formed by the length of the number line between 0 and 1.

   (c) Draw a diagonal through the square.

   (d) Recall the Pythagorean relationship. If the side length of the square is 1 unit, what is the length of the diagonal?

   (e) Use the length of the diagonal to locate $\sqrt{2}$ along the number line.

   (f) Use a similar method to locate $\sqrt{3}$, $\sqrt{5}$, and $\sqrt{7}$.

1.4 Solving Problems

In this lesson, you have learned how to **interpret mathematical information** when it is presented either in words or in a number sentence. Recall:
- There are many different ways to use words to describe the same mathematical operation.
- In a number sentence, the operations move from left to right. In word form, the order can vary.
- To translate a problem situation into a number sentence, start by picking out key points. Then order the key points to create a number sentence.

Example 1

Tanya's score on her most recent math test was two marks more than one hundred and thirty percent of her score on the last test. If her score on this test was eighty marks, what was her score on the last test?

Solution

Step 1: Identify the key points in point form.
- score on this test was higher than on last test
- score on this test was 2 marks more than 130% of her last score
- score on this test was 80

Step 2: Order the key points to create an equation.
130% of last test score plus 2 equals 80

Step 3: Create a number sentence. Substitute mathematical symbols for any information still expressed in words. Let t represent Tanya's last test score.

$130\% \cdot t + 2 = 80$	The word "of" indicates multiplication.
$130\% \cdot t = 78$	Subtract 2 from both sides.
$1.3t = 78$	Express 130% as a decimal.
$\dfrac{1.3t}{1.3} = \dfrac{78}{1.3}$	Divide both sides by 1.3.
$t = 60$	Simplify.

Tanya's score on the last test was 60 marks.

Step 4: Check to make sure the result fits the original problem situation.
Tanya's score on this test, 80, is 2 more than 130% of her score on the last test, 60.

Therefore, 80 − 2 should equal 130% of 60.

L.S.	R.S.
80 − 2	1.3 · 60
= 78	= 78

The left side is equal to the right side, so the answer, 60 marks, is correct.

Example 2

A mathematics teacher invited her students to perform this series of operations. No matter what starting number the students chose, the result was always 6. Why?

1. Choose any number from one to nine.
2. Add four to the number.
3. Multiply your answer by two.
4. Subtract eight.
5. Multiply by three.
6. Divide by the number you started with.

Solution

Write the series of instructions as arithmetic expressions and simplify each step as much as possible. Use n to represent the variable starting number.

1. Choose a number from 1 to 9.
   n
2. Add 4 to the number.
   $n + 4$
3. Multiply your answer by 2.
   $(n + 4) \cdot 2 = 2n + 8$
4. Subtract 8.
   $2n + 8 - 8 = 2n$
5. Multiply by 3.
   $3 \cdot (2n) = 6n$
6. Divide by the number you started with.
   $6n \div n = 6$

The result is always 6 because you always end up dividing 6 times the number by the number.

Exercises

1. Complete the table by listing at least five different words or phrases that indicate each operation.

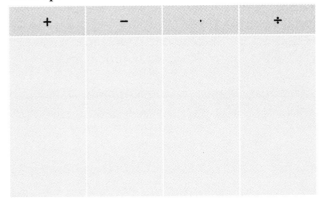

+	−	·	÷

2. Match each number statement to the correct verbal description.
   (a) one fourth of the square of six
   (b) the square of the quotient of six divided by four
   (c) six squared is one fourth of a number
   (d) one fourth of six multiplied by six squared
   (e) six squared to the fourth
   (f) the product of the square root of six squared and thirty-six fourths

   (i) $(6^2)^4$

   (ii) $6^2 = \frac{1}{4}n$

   (iii) $\sqrt{6^2 \cdot \frac{36}{4}}$

   (iv) $\frac{1}{4} \cdot 6^2$

   (v) $\left(\frac{6}{4}\right)^2$

   (vi) $\frac{6}{4} \cdot 6^2$

3. Which of the expressions in Problem 2 have the same value?

4. Express each sequence of operations in words.
   (a) $2 + 3 - 4$

   (b) $5 \cdot 6 + 8$

   (c) $\sqrt{37}$

   (d) $(22 \cdot 4^2) \div 4$

   (e) $\sqrt{5} + \sqrt{7}$

   (f) $(9 \cdot 4)(3 \cdot 5)$

   (g) $\frac{6^3}{6^2}$

   (h) $(4 - 12)^2$

   (i) $\frac{75 - (-2^5)}{2}$

   (j) $\left(\sqrt[3]{117}\right)^4$

5. Express with numbers, then simplify.

   (a) the square root of six

   (b) ten reduced by six

   (c) twelve multiplied by itself

   (d) the sum of seventeen and thirteen

   (e) the reciprocal of four

   (f) the sum of the first six natural numbers

   (g) the square root of two hundred forty-three

   (h) nine squared plus twenty-six

   (i) the average of sixty-two and seventy-three

   (j) twenty percent of six hundred

   (k) the sum of the reciprocals of six, nine, and twelve

6. Use a calculator to find each result.

   (a) $6^2 - 34^5$

   (b) $8 \cdot (5^2)^3$

   (c) $5 - \sqrt{7^8}$

   (d) $2^3 + 7^3$

   (e) $34 \cdot 6^7 \cdot 3^4$

   (f) $\dfrac{6^7}{7^9}$

   (g) $9^0 \cdot 5^5$

   (h) $\sqrt[3]{4}$

7. Describe any errors in Sachi's solution and then correct the result.

Problem

Find the difference of twelve cubed and five squared and then find forty percent of the result.

Solution

$12^3 - 5^2 \cdot 40\%$

$= 1728 - 25 \cdot 40$

$= 1728 - 1000$

$= 728$

The result is 728.

8. Write a set of instructions that will allow another student to find:

   (a) $(1 + 2) \div 3$

   (b) $9 \cdot 4 \div 3 \cdot 5$

   (c) the 5% commission on a sale of $40,200

   (d) $12^2 + 18^3$

   (e) $\left(\sqrt{27}\right)^4$

9. Write a set of instructions to explain to another student how to use a scientific calculator to find the reciprocal of the square root of a number.

10. Tony is 24 years, or 75%, older than Melissa.

    (a) Write Melissa's age as a power of 2.

    (b) How old is Tony?

11. Create a number trick that is similar to the one in Example 2. Test your trick with several classmates. Can they figure out why your trick works?

In this lesson, you have used irrational number approximations to **perform calculations with complex operations.** Recall:
- Use the same order of operations you use for rational numbers.
- Use the root keys on your calculator to find approximations for irrational numbers. Do not round until you reach the final stage of the calculation.
- Use estimation to predict results and check approximations.

Example

Find the base length and height (to one decimal place) of an equilateral triangle with area 24 cm².

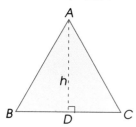

Area = 24 cm²

Solution

Begin by separating the equilateral triangle into two right triangles. Since $\triangle ABC$ is equilateral, $AB = AC = BC$.

Let the side length of the triangle be x.

The base length of each right triangle must be half of a side length or $\frac{1}{2}x$.

Area = 24 cm²

Now you can use the Pythagorean relationship to find the height in terms of x.
In $\triangle ABD$, $BD^2 + AD^2 = AB^2$.

$$BD^2 + AD^2 = AB^2$$ Substitute the known values.

$$\left(\frac{1}{2}x\right)^2 + h^2 = x^2$$ Simplify $\left(\frac{1}{2}x\right)^2 = \frac{1}{2} \cdot x \cdot \frac{1}{2} \cdot x = \frac{1}{4}x^2$.

$$\frac{1}{4}x^2 + h^2 = x^2$$ Subtract $\frac{1}{4}x^2$ from both sides.

$$h^2 = x^2 - \frac{1}{4}x^2$$ Simplify the right side.

$$h^2 = \frac{3}{4}x^2$$ Find the square root of each side.

$$h = \sqrt{\frac{3}{4}x^2}$$

$$h = \sqrt{\frac{3}{4}} \cdot \sqrt{x^2}$$ Simplify $\sqrt{x^2}$.

$$h = x\sqrt{\frac{3}{4}}$$ Express $\frac{3}{4}$ as a decimal, so you can enter it on a calculator.

$$h = x\sqrt{0.75}$$ Calculate $\sqrt{0.75}$.

$$h \approx 0.866025403x$$

To find h, you need to calculate the side length, x.
You know the area of $\triangle ABC$ is 24 cm², so you can use the area formula to calculate x.

$$A = \frac{1}{2} \cdot b \cdot h$$

$$24 = \left(\frac{1}{2}x\right)(0.866025403x)$$

$$24 = (0.5x)(0.866025403x)$$

$$24 = 0.433012701x^2$$

$$x^2 \approx 55.42562596$$

$$x \approx 7.44483888$$

$$x \approx 7.4$$

Now you know the side length of the triangle, 7.4 cm, and you can calculate the height.

$$h \approx 0.866025403x$$

$$h \approx 0.866025403(7.4)$$

$$h \approx 6.408587982$$

$$h \approx 6.4$$

The side length of the triangle is about 7.4 cm and the height is about 6.4 cm.

Check these measurements by substituting them into the area formula: $A = \frac{1}{2} \cdot b \cdot h$

L.S.	R.S.
24	$\frac{1}{2}(7.4) \cdot 6.4$
	$= 3.7 \cdot 6.4$
	$= 23.68$

The left side is approximately equal to the right side, so the measurements are very close to correct.

Exercises

1. You can estimate square roots with a number line or by using fractions. Demonstrate each procedure and compare the accuracy of the results.

2. Describe the keystroke sequence you would use on your calculator to approximate $\sqrt{15}$.

3. Show how you would round $\sqrt{15}$ to three decimal places. Explain each step in the rounding procedure.

4. Estimate to two decimal places.
   (a) $\sqrt{6}$ (b) $\sqrt{21}$ (c) $\sqrt{43}$

   (d) $2\sqrt{37}$ (e) $5\sqrt{117}$ (f) $3\sqrt{73}$

5. Describe the keystroke sequence you would use on your calculator to approximate $\sqrt[3]{15}$.

6. Calculate to three decimal places.
   (a) $\sqrt{2} + \sqrt{5}$ (b) $4\sqrt{2} - 3\sqrt{2}$

   (c) $\dfrac{30\sqrt{18}}{3\sqrt{2}}$ (d) $3\sqrt{6} - 2\sqrt{5}$

   (e) $\left(12\sqrt{5}\right)\left(-10\sqrt{7}\right)$ (f) $5\sqrt{2} + 2\sqrt{3}$

(g) $\left(2\sqrt{7}\right)\left(4\sqrt{3}\right)$

(h) $\sqrt{5}+\sqrt[3]{2}-\sqrt{3}$

(i) $\dfrac{6\sqrt{12}}{6\sqrt{3}}$

(j) $\sqrt[3]{6}-\sqrt{5}+4\sqrt{4}$

(k) $\dfrac{50\sqrt{12}}{-10\sqrt{3}}$

(l) $\left(2\sqrt{10}\right)\left(4\sqrt{5}\right)$

7. Describe, in order, the steps you will use to simplify this calculation to two decimal places. Then simplify.

$$\frac{3\sqrt{5}-2\sqrt{2}+2\sqrt{3}\cdot 4\sqrt{3}}{5\sqrt{2}}$$

8. Calculate to three decimal places.

(a) $\dfrac{2\sqrt{5}+\sqrt{5}}{\sqrt{6}}$

(b) $\dfrac{3}{5\sqrt{2}-\sqrt{2}}$

(c) $\left(3\sqrt{2}+5\sqrt{2}\right)\left(9\sqrt{2}\right)$

(d) $\dfrac{6\sqrt{3}\left(2\sqrt{8}\right)+5\sqrt{5}}{4\sqrt{2}}$

9. Correct any errors in Dane's solution.

Problem

Find the value, to two decimal places, of

$$\frac{3\sqrt{11}}{2\sqrt{5}} \cdot \sqrt{6} + 4\sqrt{2}$$

Dane's Solution

$$\frac{3\sqrt{11}}{2\sqrt{5}} \approx \frac{9.949874371}{4.472135955}$$

$$\approx 2.24859546$$

$$\sqrt{6} + 4\sqrt{2} \approx 8.106343992$$

$$2.24859546 \cdot 8.106343992 \approx 18.2278883$$

$$\approx 18.23$$

10. Janice is building a circular concrete walkway around a statue. Find the cost of the walkway if the combined cost of labor and concrete is $5/ft^2.

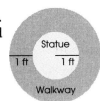

11. What length of fencing will Ray need to build a circular enclosure with a diameter of 18 ft? (Give your answer to the nearest foot.)

12. Use the Pythagorean relationship to find the sum of the perimeters of the large and small diamonds.

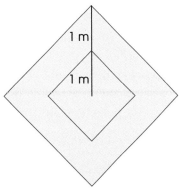

2 RADICALS

2.1 Rational Exponents

You have learned how to **convert** between radical notation and exponential notation for powers.

You have also **evaluated powers** and **applied the exponent laws** with rational exponents.

You have learned that: $x^{\frac{m}{n}} = \left(\sqrt[n]{x}\right)^m = \sqrt[n]{x^m}$ where x is a positive integer, m is an integer, and $\frac{m}{n}$ is in simplest terms.

If n is even, x must be greater than or equal to 0.

Example 1

Rewrite $64^{\frac{2}{3}}$ as a radical. Then evaluate the expression.

Solution

Method 1

$$64^{\frac{2}{3}} = \left(\sqrt[3]{64}\right)^2$$
$$= (4)^2$$
$$= 16$$

Method 2

$$64^{\frac{2}{3}} = \sqrt[3]{64^2}$$
$$= \sqrt[3]{4,096}$$
$$= 16$$

In most cases, it is easier to calculate the root before taking the power, as in Method 1.

Example 2

Evaluate the expression $\left(\sqrt[5]{x^2} \div \sqrt[3]{x}\right)^5$ where $x = 8$.

Solution

Step 1: Simplify the expression.

$$\left(\sqrt[5]{x^2} \div \sqrt[3]{x}\right)^5 = \left(x^{\frac{2}{5}} \div x^{\frac{1}{3}}\right)^5 \quad \text{Express the radicals using exponents.}$$

$$= \left(x^{\frac{6}{15}} \div x^{\frac{5}{15}}\right)^5 \quad \text{Express the rational exponents with a common denominator.}$$

$$= \left(x^{\frac{1}{15}}\right)^5 \quad \text{Divide powers by subtracting the exponents.}$$

$$= x^{\frac{5}{15}} \quad \text{Find a power of a power by multiplying the exponents.}$$

$$= x^{\frac{1}{3}} \quad \text{Express the exponent in simplest form.}$$

Step 2: Substitute 8 for x in the simplified expression.

$$x^{\frac{1}{3}} = 8^{\frac{1}{3}} = \sqrt[3]{8} = 2$$

Exercises

1. In $\sqrt[7]{9^3}$, name the root index and the radicand.

2. If you can recognize perfect squares and cubes, it can save time in determining values for square roots and cube roots. Complete as much of the table as you can without using a calculator and then calculate any remaining numbers.

Number	Number Squared	Number Cubed
1	1	1
2	4	8
3		
4		
5		

3. Write the exponential form.

   (a) $\sqrt[3]{x}$

   (b) $\sqrt[5]{3^{-2}}$

   (c) $\sqrt[6]{94}^{17}$

   (d) $\dfrac{1}{\sqrt{52^7}}$

   (e) $\sqrt[3]{13x^2}$

4. Write the radical.

   (a) $n^{\frac{1}{4}}$

   (b) $11^{-\frac{1}{2}}$

   (c) $w^{\frac{8}{3}}$

   (d) $45^{0.8}$

   (e) $\left(3b^2\right)^{\frac{1}{5}}$

5. For $\sqrt[n]{x}$, where n is even, what restrictions must be placed on x so that the value of the radical is a real number? Demonstrate using an example.

Work with a partner.

6. **(a)** What is the value of $(-27)^{\frac{2}{3}}$?

(b) Discuss the following question with your partner. Be prepared to defend your opinion to the class.

Is $(-27)^{\frac{2}{3}}$ equal to $(-27)^{\frac{4}{6}}$? Why or why not?

(Hint: Compare the denominators 3 and 6. Check the restrictions placed on roots.)

7. Evaluate. Try to make as many calculations as possible without using a calculator. Be aware of restrictions such as the one you encountered in Problem 6.

(a) $\sqrt[797]{(-9)^{797}}$

(b) $25^{\frac{3}{2}}$

(c) $(-64)^{\frac{2}{3}}$

(d) $\sqrt[5]{32^{10}}$

(e) $\left(\frac{1}{8}\right)^{-\frac{2}{3}}$

8. Simplify completely using exponent laws. Assume all variables are greater than zero.

(a) $\sqrt[3]{x^5} \cdot \sqrt[3]{x}$

(b) $\sqrt[9]{k^{\frac{2}{3}} \div k^{\frac{1}{3}}}$

9. Simplify the expression and then evaluate for $x = 64$.

$$\sqrt[3]{\frac{1}{x^2}} \cdot \sqrt{x^3}$$

10. Create a problem similar to one on this page. Write a solution for your problem on a separate piece of paper and then exchange problems with your partner. Solve the problems and compare solutions. Offer constructive suggestions about how each problem and solution might be improved.

11. Write a journal entry to describe what you have learned about radicals and exponential notation. Make a list of key vocabulary terms and definitions. Comment on the advantages and disadvantages of using the computer-guided lesson to learn about rational exponents.

2.2 Simplifying Radical Expressions

In this lesson, you learned how to simplify radical expressions using the product and quotient properties of radicals. You also learned how to add and subtract radical expressions.

Product property of radicals: If $\sqrt[n]{a}$ and $\sqrt[n]{b}$ are real numbers, then $\sqrt[n]{a}\sqrt[n]{b} = \sqrt[n]{ab}$.

Quotient property of radicals: If $\sqrt[n]{a}$ and $\sqrt[n]{b}$ are real numbers and $b \neq 0$, then= $\dfrac{\sqrt[n]{a}}{\sqrt[n]{b}} = \sqrt[n]{\dfrac{a}{b}}$.

Radical expressions with the same index and the same radicand are called like or similar radicals. To add or subtract radicals, simplify each radical expression and combine all like radicals by adding or subtracting the coefficients and keeping the common radical.

A radical expression is in simplest form when:
1. No radicals appear in the denominator of a fraction.
2. The radicand contains no fractions or negative numbers.
3. Each factor in the radicand is a power with an exponent that is less than the index of the radical.

Example 1

Simplify $\dfrac{\sqrt{32x^5}}{\sqrt{2x}}$.

Solution

❶ $\dfrac{\sqrt{32x^5}}{\sqrt{2x}} = \sqrt{\dfrac{32x^5}{2x}}$

❷ $\qquad = \sqrt{16x^4}$

❸ $\qquad = 4x^2$

❶ Apply the quotient property of radicals.

❷ $\dfrac{32}{2} = 16$ and $\dfrac{x^5}{x} = x^4$

❸ The square root of 16 is 4, and the square root of x^4 is x^2.

Example 2

Simplify $\sqrt[3]{24x^4}$.

Solution

$24x^4$ is not a perfect cube, so find the largest perfect cube factor of $24x^4$.

❶ $\sqrt[3]{24x^4} = \sqrt[3]{(8)(3)(x^3)(x)}$

❷ $\qquad = \sqrt[3]{(8x^3)(3x)}$

❸ $\qquad = \left(\sqrt[3]{8x^3}\right)\left(\sqrt[3]{3x}\right)$

❹ $\qquad = 2x\sqrt[3]{3x}$

❶ 24 can be rewritten as (8)(3) since 8 is a

perfect cube, and x^4 can be rewritten as $(x^3)(x)$ since x^3 is a perfect cube.

❷ Combine the perfect cubes and combine the other factors.

❸ Apply the product property of radicals.

❹ The cube root of 8 is 2, and the cube root of x^3 is x.

Example 3

Simplify $\sqrt[3]{54} - 3\sqrt[3]{16} + 4\sqrt[3]{128}$, and then combine like radicals.

Solution

To simplify $\sqrt[3]{54} - 3\sqrt[3]{16} + 4\sqrt[3]{128}$, first create like radicals. Find the largest perfect cube factors of 54, 16, and 128.

$$\sqrt[3]{54} - 3\sqrt[3]{16} + 4\sqrt[3]{128}$$

❶ $= \sqrt[3]{(27)(2)} - 3\sqrt[3]{(8)(2)} + 4\sqrt[3]{(64)(2)}$

❷ $= \left(\sqrt[3]{27}\right)\left(\sqrt[3]{2}\right) - 3\left(\sqrt[3]{8}\right)\left(\sqrt[3]{2}\right) + 4\left(\sqrt[3]{64}\right)\left(\sqrt[3]{2}\right)$

❸ $= 3\sqrt[3]{2} - (3)(2)\sqrt[3]{2} + (4)(4)\sqrt[3]{2}$

❹ $= 3\sqrt[3]{2} - 6\sqrt[3]{2} + 16\sqrt[3]{2}$

❺ $= 13\sqrt[3]{2}$

❶ The largest perfect cube factors of 54, 16, and 128, are 27, 8, and 64, respectively.

❷ Apply the product property of radicals.

❸ Find the cube roots of 27, 8, and 64.

❹ (3)(2) = 6 and (4)(4) = 16

❺ 3 − 6 + 16 = 13

Exercises

1. Define each term, using examples.
   (a) radicand

   (b) index

   (c) mixed radicals

2. Complete each statement.
   (a) To _____ $\sqrt{24}$ means to write it as $2\sqrt{6}$.

   (b) The largest perfect square _____ of 27 is 9.

   (c) Numbers such as 1, 4, 9, 16, 25, and 36 are called perfect _____.

   Numbers such as 1, 8, 27, 64, and 125 are called perfect _____.

3. Consider the expressions $\sqrt{4 \cdot 5}$ and $\sqrt{4} \cdot \sqrt{5}$.
   (a) Which expression is the square root of a product?

   (b) Which expression is product of square roots?

   (c) How are the two expressions related?

4. Explain, using an example, how each property of radicals can be used.
   (a) product property

   (b) quotient property

5. Explain what is meant by *like radicals*. Give three examples of radical terms that are similar and three examples of radical terms that are not similar. Include exponents and variables in your examples.

6. Write two radical expressions:

    (a) that have the same radicand but a different index.

    (b) that have the same index but a different radicand.

7. Can the expressions in Problem 6(a) be added? Why or why not? Can the expressions in (b) be added? Why or why not?

8. Find and correct the error in the solution.

Problem

Simplify $\sqrt[3]{250}$.

Solution

$\sqrt[3]{250}$
$= \sqrt[3]{(125)+(125)}$
$= \sqrt[3]{125} + \sqrt[3]{125}$
$= 5 + 5$
$= 10$

9. Simplify.

    (a) $\sqrt{240}$ **(b)** $\sqrt[4]{32}$ **(c)** $\sqrt[3]{270}$

    (d) $\sqrt[3]{-81}$ **(e)** $-2\sqrt[5]{-96}$ **(f)** $\sqrt{\dfrac{18}{100}}$

    (g) $\sqrt[3]{\dfrac{-16}{27}}$

10. Simplify. Assume that all variables represent positive real numbers.

    (a) $\sqrt{8x^5}$ **(b)** $\sqrt[3]{r^{17}}$ **(c)** $\sqrt[3]{16x^5y^4}$

    (d) $3\sqrt[3]{27j^7k}$ **(e)** $\sqrt{\dfrac{17xy}{64a^4}}$ **(f)** $\sqrt[3]{\dfrac{-54x^6}{125x^3}}$

(g) $\sqrt{\dfrac{200x^2y}{49x^4}}$

11. Simplify.

    (a) $(\sqrt{11})(\sqrt{11})$

    (b) $\left(\sqrt[3]{7x}\right)\left(\sqrt[3]{49x^2}\right)$

    (c) $\left(\sqrt[4]{2x^3}\right)\left(\sqrt[4]{8x}\right)$

    (d) $\dfrac{\sqrt{32}}{\sqrt{8}}$

    (e) $\dfrac{\sqrt[3]{64}}{\sqrt[3]{8}}$

    (f) $\dfrac{\sqrt[3]{54x^5}}{\sqrt[3]{2x^2}}$

12. Simplify and combine like radicals.

    (a) $\sqrt{8}-\sqrt{2}$

    (b) $\sqrt{125}+\sqrt{25}$

    (c) $\sqrt{72}-\sqrt{32}$

    (d) $\sqrt[3]{80}-\sqrt[3]{10,000}$

    (e) $3\sqrt[4]{512}+2\sqrt[4]{32}$

    (f) $4\sqrt[4]{243}-\sqrt[4]{48}$

    (g) $\sqrt{98}-\sqrt{50}-\sqrt{72}$

    (h) $\sqrt{20}+\sqrt{125}-\sqrt{80}$

    (i) $\sqrt{32x^3}+\sqrt{50x^3}-\sqrt{18x^3}$

    (j) $\sqrt[3]{16x^4}+\sqrt[3]{54x^4}$

    (k) $3\sqrt{2x}-\sqrt{8x}$

    (l) $\sqrt{25y^2z}-\sqrt{16y^2z}$

13. A rectangular room measures $\sqrt{18}$ yd by $\sqrt{32}$ yd. How much carpet is needed to carpet the room?

Multiplying and Dividing Radical Expressions

In this lesson, you learned how to multiply and divide radical expressions, and to express the results in simplified form. Recall:

- To multiply two radicals, apply the **product property of radicals**.

For real numbers a and b and integer $n > 1$, $\sqrt[n]{a} \cdot \sqrt[n]{b} = \sqrt[n]{ab}$
If n is even, a and b must be ≥ 0.

- To divide two radicals, first rationalize the denominator to eliminate any radicals and then apply the **quotient property of radicals**.

For real numbers a and b, $b \neq 0$, and integer $n > 1$, $\dfrac{\sqrt[n]{a}}{\sqrt[n]{b}} = \sqrt[n]{\dfrac{a}{b}}$
If n is even, a must be ≥ 0 and b must be > 0.

Example 1

Find the product. Simplify if possible.

$$3\left(\sqrt{5x} - \sqrt{3}\right)^2$$

Solution

Write the squared binomial as a product of two binomials.

$$3\left(\sqrt{5x} - \sqrt{3}\right)^2$$
$$= 3\left(\sqrt{5x} - \sqrt{3}\right)\left(\sqrt{5x} - \sqrt{3}\right)$$

Use the FOIL method to multiply the binomials.

$$= 3\left(\sqrt{5x}\sqrt{5x} - \sqrt{5x}\sqrt{3} - \sqrt{3}\sqrt{5x} + \sqrt{3}\sqrt{3}\right)$$

Use the product property to multiply each pair of radicals.

$$= 3\left(\sqrt{25x^2} - \sqrt{15x} - \sqrt{15x} + \sqrt{9}\right)$$

Simply the radicals wherever possible.

$$= 3\left(5x - \sqrt{15x} - \sqrt{15x} + 3\right)$$

Combine like terms.

$$= 3\left(5x - 2\sqrt{15x} + 3\right)$$

Apply the distributive property to eliminate the brackets.

$$= 3 \cdot 5x - 3 \cdot 2\sqrt{15x} + 3 \cdot 3$$

Simplify the products.

$$= 15x - 6\sqrt{15x} + 9$$

Example 2

Simplify by rationalizing the denominator. $\dfrac{2}{\sqrt[3]{6}}$

Solution

Multiply the numerator and denominator by a factor that will result in a perfect cube in the denominator.

$$\frac{2}{\sqrt[3]{6}} = \frac{2 \cdot \sqrt[3]{6 \cdot 6}}{\sqrt[3]{6} \cdot \sqrt[3]{6 \cdot 6}}$$

Use the product property to multiply the radicals in the denominator.

$$= \frac{2 \cdot \sqrt[3]{6 \cdot 6}}{\sqrt[3]{6 \cdot 6 \cdot 6}}$$

Simplify the numerator and denominator.

$$= \frac{2 \cdot \sqrt[3]{36}}{6}$$

Reduce the fraction to simplest terms.

$$= \frac{\sqrt[3]{36}}{3}$$

Exercises

1. Complete each statement.

   (a) To find the product of $3\sqrt{2}$ and $5\sqrt{7}$, multiply $3 \cdot \square$ and $\sqrt{2} \cdot \square$.

   (b) To find the product of $3\sqrt{5}$ and $\left(\sqrt{2} - 2\sqrt{6}\right)$ use the $\boxed{}$ property.

   (c) To rationalize the denominator in the fraction $\dfrac{2\sqrt{2}}{\sqrt{3}}$, multiply the numerator and denominator by $\square$.

2. Which expressions can be simplified? Write the simplified form.

   (a) $4\sqrt{6} \cdot 2\sqrt{6}$

   (b) $4\sqrt{6} + 2\sqrt{6}$

   (c) $3\sqrt{2} - 2\sqrt{3}$

   (d) $3\sqrt{2} \cdot 2\sqrt{3}$

3. Which expression in each pair can be simplified? Write the simplified form.

   (a) $5 + 6\sqrt[3]{6}$ or $5 \cdot 6\sqrt[3]{6}$

   (b) $\dfrac{30\sqrt[3]{15}}{5}$ or $\dfrac{\sqrt[3]{15}}{5}$

4. Consider $\dfrac{\sqrt{3}}{\sqrt{7}} = \dfrac{\sqrt{3}\sqrt{7}}{\sqrt{7}\sqrt{7}}$. Explain why the expression on the left side is equal to the one on the right side.

5. Explain why $\dfrac{\sqrt[3]{12}}{\sqrt[3]{5}}$ is not in simplified form.

6. Explain why $\sqrt{m} \cdot \sqrt{m} = m$ but $\sqrt[3]{m} \cdot \sqrt[3]{m} \neq m$. (Assume that $m > 0$.)

7. To rationalize the denominator of $\dfrac{\sqrt[4]{2}}{\sqrt[4]{3}}$, why would you not multiply the fraction

by $\dfrac{\sqrt[4]{3}}{\sqrt[4]{3}}$?

8. Explain why $\sqrt{\dfrac{3a}{11y}}$ is not in simplified form.

9. Fill in the blanks to complete each stage in the solution.

(a)
$$5\sqrt{8} \cdot 7\sqrt{6} = 5(7)\sqrt{8\,\boxed{}}$$
$$= 35\sqrt{\boxed{}}$$
$$= 35\sqrt{\boxed{} \cdot 3}$$
$$= 35 \cdot \boxed{}\,\sqrt{3}$$
$$= \boxed{}\,\sqrt{3}$$

(b)
$$\dfrac{9}{\sqrt[3]{4a^2}} = \dfrac{9 \cdot \sqrt[3]{2a}}{\sqrt[3]{4a^2 \cdot \boxed{}}}$$
$$= \dfrac{9\sqrt[3]{2a}}{\sqrt[3]{\boxed{}}}$$
$$= \dfrac{9\sqrt[3]{2a}}{\boxed{}}$$

10. Multiply.

(a) $\sqrt[3]{8} \cdot \sqrt[3]{16}$

(b) $3\sqrt{5xy} \cdot 2\sqrt{x^3 y}$

(c) $2\sqrt{3}\left(7 - \sqrt{4}\right)$

(d) $-5\sqrt{2b}\left(2\sqrt{8b^2} - \sqrt{b^2}\right)$

11. Explain why the product of $\sqrt{m}+3$ and $\sqrt{m}-3$ does not contain a radical.

12. Give the conjugate of each expression.

(a) $2+\sqrt{6}$

(b) $3\sqrt{11}-8\sqrt{16}$

(c) $-\sqrt{35}+5$

(d) $7-\sqrt{21x}$

13. Rationalize each denominator.

(a) $\dfrac{\sqrt{5}}{\sqrt{6}}$

(b) $\dfrac{8-\sqrt{10}}{\sqrt{7}-3\sqrt{3}}$

(c) $\dfrac{5\sqrt[3]{12n}}{14\sqrt[3]{10n^5}}$

(d) $\dfrac{3}{\sqrt[4]{81x^6}}$

14. The orbital radius, (in meters), of an Earth satellite is given by $r = \sqrt[3]{\dfrac{GMt^2}{4\pi^2}}$, where G is the universal gravitational constant, M is the mass of the Earth, and t is the orbital period in seconds. Express the radical in simplified form.

3 LINEAR EQUATIONS

3.1 Simplifying Algebraic Expressions

In this lesson, you reviewed some basic properties of real numbers and learned to simplify algebraic expressions by grouping like terms and by using the distributive property. Recall:

- The **additive identity** is 0 because $0 + a = a + 0 = a$.
- The **multiplicative identity** is 1 because $1(a) = a(1) = a$.
- The product of 0 and any real number is 0: $a(0) = 0(a) = 0$
- If the sum of two real numbers is 0, they are **additive inverses**:
  $a + (-a) = (-a) + a = 0$
- If the product of two real numbers is 1, they are **reciprocals**: $a\left(\dfrac{1}{a}\right) = \left(\dfrac{1}{a}\right)a = 1$

- For basic division:

  $$\dfrac{a}{1} = a \qquad \dfrac{a}{a} = 1, \text{ if } a \neq 0 \qquad \dfrac{0}{a} = 0, \text{ if } a \neq 0 \qquad \dfrac{a}{0} \text{ is undefined}$$

- The basic division properties, for real numbers, are:
  - **commutative property**: $a + b = b + a$ and $ab = ba$
  - **associative property**: $(a + b) + c = a + (b + c)$ and $(ab)c = a(bc)$
  - **distributive property**: $a(b + c) = ab + ac$ and $a(b - c) = ab - ac$

Example 1

Simplify $-5(3m)(-4n)$.

Solution

$-5(3m)(-4n) = (-5)(3)(-4)(m)(n)$	Group the numbers and variables using the associative and commutative properties.
$= (60)(m)(n)$	Multiply the numbers: $-5 \cdot 3 \cdot (-4) = 60$
$= (60)(mn)$	Multiply the variables: $m \cdot n = mn$
$= 60mn$	Multiply the terms in parentheses.

Example 2

Simplify $4(3y - z + 10)$.

Solution

$4(3y - z + 10) = (4 \cdot 3y) - (4 \cdot z) + (4 \cdot 10)$	Use the distributive property to distribute the 4.
$= 12y - 4z + 40$	Multiply the terms inside the parentheses.

Example 3

Simplify $3x^3 + 2x^3 - 3x^2 + 4x^3$.

Solution

$3x^3 + 2x^3 - 3x^2 + 4x^3 = 3x^3 + 2x^3 + 4x^3 - 3x^2$ Use the commutative property to group like terms. $3x^3$, $2x^3$, and $4x^3$ are like terms because they share the same variable and exponent.

$$= 9x^3 - 3x^2$$ Combine the like terms.

Exercises

1. Define each term and give an example.
   (a) algebraic term
   (b) algebraic expression

   (c) variable
   (d) coefficient

   (e) constant

2. Explain when each property is used. Include an example with your explanation.
   (a) associative property

   (b) commutative property

   (c) distributive property

3. Use an example to show that each property does not apply to subtraction.
   (a) associative property
   (b) commutative property

4. Define **like terms**. Give three sets of like terms and three sets of unlike terms. Include some examples with exponents.

5. Explain why the distributive property does not apply when you are simplifying $6(2 \cdot b)$.

6. Consider the expression $2x^2 - x + 6$.
   (a) How many terms are in the expression?

   (b) What are the terms?

   (c) Give the coefficient of each term.

   (d) Are there any like terms? If so, identify them.

7. Label each pair of terms **like** or **unlike**.
   (a) $-3x, 6x$ **(b)** $5y^2, 7xy$ **(c)** $xy, -3yx$

   (d) $-3t^2, 12t^2$ **(e)** $-5xy, -5yz$

8. Simplify.
   (a) $9(-8q)$ **(b)** $(-4c)(-2d)$ **(c)** $-6(2b)$

   (d) $(3s)(7t)$ **(e)** $(2a)(3b)(-4c)$

9. In the expression $-(2x - 4)$, a negative sign precedes the parentheses.
   (a) What does this negative sign mean?

   (b) Expand and simplify $-(2x - 4)$.

10. Simplify by eliminating the parentheses.
   (a) $3(a + 2)$ **(b)** $-9(2x + y)$ **(c)** $-3(-5 + a)$

   (d) $-(-4 + z)$ **(e)** $6\left(\dfrac{2}{3}x + \dfrac{7}{6}y - \dfrac{1}{3}\right)$

11. Simplify by combining like terms.

    (a) $2a + 7a - 3a$ **(b)** $9ab + 3ab - ab$ **(c)** $3a + 7a - 8 + 2$

    (d) $12a^2 + 3a^2 - 12a$ **(e)** $6x + 10 - 6x + 2$

12. Simplify.

    (a) $6(x + 5) - 2x$ **(b)** $8(2p + 2) + 2(p - 6)$ **(c)** $2(a + 6) - 5(a + 1)$

    (d) $3x^2 + 6(x^2 - 2x) + 10x$ **(e)** $3p - (2k + 4p) + k$

13. The diagram shows the dimensions of a swimming pool.

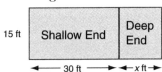

    (a) Express the area of the entire pool as the product of its length and width.

    (b) Express the area of the entire pool as the sum of the areas of the shallow and the deep ends.

    (c) Write an equation that shows that the answers to parts (a) and (b) are equal.

    (d) What property of real numbers is illustrated by your equation from part (c)?

3.2 Solving Linear Equations and Formulas

You learned how to solve equations of these forms:
- $ax + b = c$, where a, b, and c are real numbers
- $a(bx + c) = d$, where a, b, c, and d are real numbers
- $\dfrac{a}{b}x = c$, where a and b are integers and $b \neq 0$

To solve an equation, rewrite it as a series of equivalent equations until the variable is isolated and has a coefficient of 1. To create equivalent equations, you can:
- perform the same operations on both sides of the equation.
- eliminate fractions in equations by multiplying each term on both sides of the equation by the lowest common denominator.
- eliminate parentheses by applying the distributive property.

Example 1

Solve.

$$\frac{8(x-5)}{3} = 2(x-4)$$

Solution

$$\frac{8(x-5)}{3} = 2(x-4)$$
$$3\left(\frac{8(x-5)}{3}\right) = 3[2(x-4)]$$
$$8(x-5) = 6(x-4)$$
$$8x - 40 = 6x - 24$$
$$8x - 6x - 40 = 6x - 6x - 24$$
$$2x - 40 = -24$$
$$2x - 40 + 40 = -24 + 40$$
$$2x = 16$$
$$\frac{2x}{2} = \frac{16}{2}$$
$$x = 8$$

Verify the solution by substituting into the original equation.

L.S.	R.S.
$\dfrac{8(x-5)}{3}$	$2(x-4)$
$= \dfrac{8(8-5)}{3}$	$= 2(8-4)$
$= \dfrac{8(3)}{3}$	$= 2(4)$
$= \dfrac{24}{3}$	$= 8$
$= 8$	

Since 8 makes the original equation true, it verifies the solution.

Example 2

Solve each formula for the variable indicated.
(a) $V = \pi r^2 h$ for h
(b) $\dfrac{T}{6} = \dfrac{1}{6}ab(x+y)$ for x

Solution

(a)
$$V = \pi r^2 h$$
$$\frac{V}{\pi r^2} = \frac{\pi r^2 h}{\pi r^2}$$
$$\frac{V}{\pi r^2} = h$$

(b)
$$\frac{T}{6} = \frac{1}{6}ab(x+y)$$
$$6\left(\frac{T}{6}\right) = 6\left(\frac{1}{6}ab(x+y)\right)$$
$$T = ab(x+y)$$
$$T = abx + aby$$
$$T - aby = abx + aby - aby$$
$$T - aby = abx$$
$$\frac{T - aby}{ab} = \frac{abx}{ab}$$
$$\frac{T - aby}{ab} = x$$

Exercises

1. Copy and complete.
   (a) If a, b, and c are real numbers and $a = b$,
   then $a + c = b + \square$ and $a - c = b - \square$.

   (b) If a, b, and c are real numbers and $a = b$,
   then $ca = \square\, b$ and $\dfrac{a}{c} = \dfrac{\square}{}$ (for $c \neq 0$).

2. (a) In $-5b + 3 = -18$, what can you do to both sides to "undo" the operation of addition of 3?

   (b) In $-5b = -18$, what can you do to both sides to "undo" the multiplication by -5?

   (c) When solving $\dfrac{x+1}{3} - \dfrac{2}{15} = \dfrac{x-1}{5}$, why should you start by multiplying both sides by 15?

3. Determine whether -6 is a solution to each equation.
   (a) $6 - x = 2x + 24$

   (b) $\dfrac{5}{3}(x - 3) = -12$

4. Solve each equation.
   (a) $\dfrac{x}{5} = -45$

   (b) $t - 3.67 = 4.23$

   (c) $0.0035 = 0.25g$

   (d) $0 = x + 4$

   (e) $\dfrac{x}{4} = 7$

   (f) $-\dfrac{x}{6} = 8$

   (g) $1.6a = 4.032$

   (h) $0.52 = 0.05y$

5. Solve each equation.

(a) $-5x + 10 = -5(x - 2)$

(b) $-3x - 7 + x = 6x + 20 - 5x$

(c) $\dfrac{3y}{4} - 14 = -\dfrac{y}{3} - 1$

(d) $-k = -0.06$

(e) $\dfrac{5}{4}p = -10$

(f) $-\dfrac{4}{5}s = 16$

(g) $-3 = -\dfrac{9}{8}s$

(h) $8 - x = 12 - x$

(i) $8k - 2 = 13$

(j) $4a - 22 - a = -a - 7$

(k) $a + 18 = 5a - 3 + a$

6. Solve. Remember to use the distributive property where necessary.

(a) $2(2x + 1) = x + 15 + 2x$

(b) $-2(x + 5) = x + 30 - 2x$

(c) $8(3a - 5) - 4(2a + 3) = 12$

(d) $9(x + 2) = -6(4 - x)$

(e) $3(x + 2) - 2 = -(5 + x) + x$

(f) $4(y - 1) = 28$

7. Solve each formula for the indicated variable.

(a) $E = mc^2$ for m

(b) $T - w = ma$ for w

(c) $F = \dfrac{9}{5}C + 32$ for C

(d) $K = \dfrac{Mv^2}{2} + \dfrac{Iw^2}{2}$ for I

8. Debbie solved the formula $K = Ax + AB$ for B. Find the error in her solution and correct it.

$$K = Ax + AB$$
$$K - Ax = AB$$
$$\frac{K - Ax}{A} = \frac{AB}{A}$$
$$\frac{K - \cancel{Ax}}{\cancel{A}} = \frac{\cancel{A}B}{\cancel{A}}$$

9. The cost of electricity in a certain city is given by the formula $C = 0.07n + 6.50$, where C is the cost and n is the number of kilowatt hours used. Find the number of kilowatt hours used each month by a homeowner who paid $49.97.

10. A monthly water bill in a certain city is calculated using the formula

$n = \dfrac{5{,}000C - 17{,}500}{6}$ where n is the number of gallons used and C is the monthly cost. Find the cost to use 1,200 gallons of water.

11. Create an equation to fit each of these forms. In each equation, the values for a, b, and c should be rational numbers.

(a) $ax + b = c$

(b) $a(bx + c) = d$

(c) $\dfrac{a}{b}x = c$

Write a complete solution for each of your equations on another piece of paper. Then trade equations with a classmate. Check the solutions by comparing your results.

3.3 Applications of Equations

In this lesson, you learned how to solve problems involving diagrams, percents, mixtures, and formulas. You examined three methods for solving problems:
- translating the text of the problem into an equation
- using a diagram to help visualize a problem before forming an equation
- using a formula suggested by the information in a problem

You solved the problems by following these steps:
1. Think about the problem.
2. Make a plan.
3. Solve the problem.
4. Look back.
5. Look ahead.

Example 1

A wedding gown that normally sells for $397.98 is on sale for $265.32. Find the percent of markdown.

Solution

Step 1: Think about the problem.
Given information
The regular selling price is $397.98.

The sale price is $265.32.

Needed information
the percent of markdown

Step 2: Make a plan.
The markdown is the product of the regular selling price and the percent of markdown. The sale price is the regular selling price minus the markdown. Let m be the percent of markdown.

Step 3: Solve the problem.

$$265.32 = 397.98 - \frac{m}{100} \cdot 397.98$$

❶ $\quad\quad 26{,}532 = 39{,}798 - 397.98m$
❷ $\quad 26{,}532 + 397.98m = 39{,}798$
❸ $\quad\quad 397.98m = 39{,}798 - 26{,}532$
❹ $\quad\quad \dfrac{397.98m}{397.98} = \dfrac{13{,}266}{397.98}$
$\quad\quad\quad\quad m = 33.333...$

❶ Multiply both sides by 100 to eliminate the denominator.
❷ Add $397.98m$ to both sides.
❸ Subtract 26,532 from both sides.
❹ Divide both sides by 397.98.
Therefore, the percent of markdown is $33\frac{1}{3}\%$.

Step 4: Look back.
The markdown is $33\frac{1}{3}\%$ of $397.98, or $132.66.

The sale price is $397.98 – $132.66, or $265.32.

The solution is correct.

Example 2

Two runners start a race at the same time, one running 12 mph and the other 10 mph. If they each maintain their pace, when will they be one-quarter of a mile apart?

Solution

Step 1: Think about the problem.
Given information
One runner runs at a speed of 12 mph and the other runner runs at a speed of 10 mph.

Needed information
the time it will take them to be one-quarter of a mile apart

Step 2: Make a plan.
Use the formula *distance = speed · time.*

The distance covered by the faster runner will be one-quarter of a mile more than the distance covered by the slower runner.

faster speed · time = slower speed · time + 0.25

Step 3: Solve the problem.

$12 \cdot t = 10 \cdot t + 0.25$
$12t = 10t + 0.25$
$12t - 10t = 0.25$ Subtract $10t$ from both sides.
$2t = 0.25$ Simplify.
$2t \div 2 = 0.25 \div 2$ Divide both sides by 2.
$t = 0.125$

Therefore, the runners will be one-quarter of a mile apart after 0.125 h or 7.5 min.

Step 4: Look back.

In one hour, the two runners would be 2 mi apart.

They would be 1 mi apart in 30 min.

They would be one-quarter of a mile apart in 30 ÷ 4 = 7.5 min.

The solution is correct.

Exercises

1. Complete each statement.
   (a) If the sum of the measures of two angles is 90°, the angles are called _____ angles.

   (b) If a triangle has two sides with equal measures, it is called an _____ triangle.

   (c) When an investment is made, the amount of money invested is called the _____.

   (d) The _____ of several values is the sum of those values divided by the number of values.

   (e) When the regular price of an item is reduced, the amount of reduction is called the _____.

2. The intensity of sound is measured in *decibels*. Translate the descriptions in the right column into mathematical symbols to complete the decibel column.

Source of Noise	Decibels (d)	Compared to Conversation
Conversation	d	
Vacuum cleaner		15 decibels more
Circular saw		10 decibels less than twice
Jet takeoff		20 decibels more than twice
Whispering		10 decibels less than half
Rock band		twice

3. The table shows the four types of problems an instructor put on a history test.
   (a) Complete the table.

Type of Question	Number	Value of Each Question
Multiple choice	x	5
True/false	3x	2
Essay	x – 2	10
Fill-in	x	5

   (b) Which type of question appears most frequently on the test?

   (c) Write an algebraic expression that represents the total number of points on the test.

4. If $x of $30,000 is invested at 5%, how much is left to be invested at another rate?

5. **(a)** Complete each solution.

$$0.09x + 0.08(2{,}000 - x) = 400$$

$$\boxed{}[0.09x + 0.08(2{,}000 - x)] = \boxed{}(400)$$

$$\boxed{} + 8(2{,}000 - x) = \boxed{}$$

$$9x + \boxed{} - 8x = 40{,}000$$

$$x = \boxed{}$$

(b)

$$0.2(5) + 0.6x = 0.4(5 + x)$$

$$\boxed{}[0.2(5) + 0.6x] = \boxed{}[0.4(5 + x)]$$

$$\boxed{}(5) + 6x = \boxed{}(5 + x)$$

$$10 + 6x = 20 + \boxed{}$$

$$\boxed{} = 10$$

$$x = \boxed{}$$

6. Briefly explain what should be accomplished in each step of the problem-solving strategy. (The steps are listed in the summary.)

7. If a car travels 60 mph for 30 minutes, explain why the distance traveled is not $60 \cdot 30 = 1{,}800$ miles. How far does the car travel?

8. Can a mixture made from solutions with concentrations of 12% and 30% have a concentration less than 12% or greater than 30%? Why?

9. The sum of Jakob's age and David's age is 34 years. Five years ago, the sum of twice Jakob's age and three times David's age was 61 years. What are their present ages?

10. A chemist mixes a 30% sugar solution and a 40% sugar solution to obtain 50 gal of a 38% solution. How much of each solution was used?

11. A bank robber runs away from a robbery at 12 ft/s. One minute later, a police officer leaves the bank and pursues the robber at 16 ft/s. How long will it take the police officer to catch up with the robber?

12. Last year, Mrs. Sansom invested part of her $10,000 in savings in NewTekniks and the rest of it in a Certificate of Deposit. If she earned $1,341 in interest, how much did she invest at each rate?
- **A** NewTekniks, software developer
  High yield: 15% per year
- **B** Certificate of Deposit (CD)
  Low yield: 4.4% per year

13. Maurice's garden has a path of uniform width all around it. He put a decorative border around the outside of the path. How wide is the walkway if the border is 60 ft long?

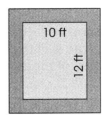

14. Create a problem that requires the use of each strategy, then exchange with a classmate.
- **(a)** translating phrases from English into an equation

- **(b)** interpreting the information on a diagram

- **(c)** substituting data into a formula

In this lesson, you learned about the axes and the quadrants of the coordinate plane. The location of any point on the coordinate plane can be stated using an ordered pair of numbers. In an ordered pair:

- The first number, the *x*-coordinate, represents the point's horizontal distance and direction along the *x*-axis in relation to the origin.
- The second number, the *y*-coordinate, represents the point's vertical distance and direction along the *y*-axis in relation to the origin.

You also learned how to convert data from a table or chart into ordered pairs. Once these ordered pairs are plotted on the coordinate system, the graph can be used to answer questions about the data.

Example

The distance a skier lands from the end of a ski jump depends on the starting height of the jump. Graph the data to answer the questions.

(a) How far from the end of the jump would a skier land if she started from an initial height of 20 yd?

(b) From what height would a skier have to start to jump a distance of 60 yd?

Starting Height (yd)	Jumping Distance (yd)
5	10
15	30
25	50
35	70

Solution

Since the jumping distance depends on the starting height, jumping distance is the dependent variable and is plotted along the *y*-axis. If each ordered pair represents (*starting height, jumping distance*), the pairs are: (5, 10), (15, 30), (25, 50), and (35, 70).

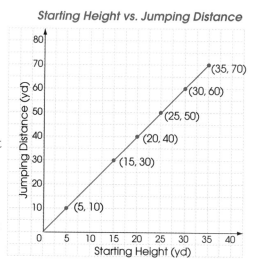

Starting Height vs. Jumping Distance

(a) A skier who started at a height of 20 yd would jump 40 yd. This is determined by moving across to 20 on the *x*-axis, then up to the point where the vertical line for 20 meets the graph (at 40 on the *y*-axis).

(b) A skier would need a starting height of 30 yd to jump 60 yd. This is determined by moving up to 60 on the *y*-axis, then across to the point where the horizontal line for 60 meets the graph (at 30 on the *x*-axis).

Exercises

1. Choose any term and explain how it relates to another term. Then relate the second term to a third. Continue until you have used all the terms. Relate the last term to the one you started with. Compare your results with a classmate's.

   origin   *ordered pair*   *quadrant*
   y-coordinate   *coordinate plane*   *x-coordinate*
   y-axis   *x-axis*   *point*
   horizontal line   *vertical line*   *intersection*

2. State the coordinates of each point.

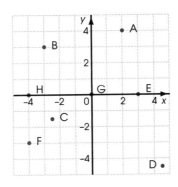

3. Plot each point on a coordinate plane.
   (a) $A(4, 3)$ (b) $B(-2, 1)$ (c) $C(3.5, -2)$
   (d) $D(-2.5, -3)$ (e) $E(5, 0)$ (f) $F(-4, 0)$

   (g) $G\left(\dfrac{8}{3}, 0\right)$ (h) $H\left(0, \dfrac{10}{3}\right)$

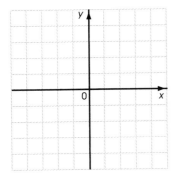

4. Plot each point. Then draw a line to connect the points.

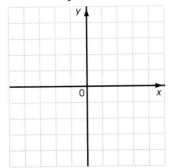

x	y
-3	-4
-1	-2
1	0
3	2
4	3

5. Use the graph to answer each question.

    (a) When did petroleum imports decline?

    (b) When did petroleum production increase?

    (c) When did imports surpass production?

    (d) Estimate the difference in imports and production for 1996.

6. A coordinate system that designates the location of places on Earth's surface uses a series of latitude and longitude lines. Longitude is listed first.

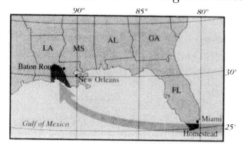

    (a) Express the coordinates of New Orleans as an ordered pair.

    (b) In August of 1992, Hurricane Andrew destroyed Homestead, Florida. Estimate the coordinates of Homestead.

    (c) Estimate the coordinates of where the hurricane entered Louisiana.

7. The map shows the area where damage was caused by an earthquake.

    (a) Find the coordinates of the epicenter (the source of the quake).

    (b) Was damage done at point $(4, 5)$?

    (c) Was damage done at point $(-1, -4)$?

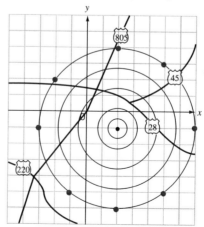

8. The diagram shows a cross-sectional profile of the Sierra Nevada mountain range in California.

(a) Estimate the coordinates of blue oak, sagebrush scrub, and tundra, using an ordered pair of the form (*distance, elevation*).

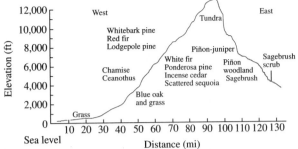

(b) The treeline is the highest elevation at which trees grow. Estimate the treeline for this mountain range.

9. The thawing guidelines that appear on the label of a frozen turkey are listed in the table. Draw a step graph that illustrates these instructions.

Size	Thawing Time
10 lb to just under 18 lb	3 days
18 lb to just under 22 lb	4 days
22 lb to just under 24 lb	5 days
24 lb to just under 30 lb	6 days

10. A train travels from Alevena to Fargo and makes these stops:

(a) Graph the data.

(b) Between which two towns was the train traveling fastest?

(c) Between which two towns was the train traveling slowest?

Town	Distance (miles)	Time (minutes)
Alevena	0	0
Biggar	100	75
Chelan	175	150
Doremy	325	225
Ethelberg	400	275
Fargo	525	375

11. Draw a simple design on a coordinate grid so each vertex of the design is at a point where grid lines intersect. Label each vertex with a letter and an ordered pair. Sit back-to-back with a partner. Describe your design and have your partner plot each point on a coordinate grid. Compare your partner's results with your own. Then repeat with your partner's design.

POLYNOMIALS AND FACTORING

4.1 Multiplying Polynomials

You have learned how to **multiply polynomials** using the distributive property.
- **Predict** the number of partial products.
- **Multiply** each term of the first polynomial by each term of the second polynomial.
- **Simplify** the resulting polynomial by **collecting like terms**.
- **Arrange** the terms of the expanded polynomial in descending order.

If more than two polynomials are to be multiplied, you can use a similar procedure. Start by finding the product of the first two and then multiply the result by the next polynomial, and so on.

Example 1

Expand $(x + 2)(x - 3)(x^2 - 4)$.

Solution

❶ $[(x + 2)(x - 3)](x^2 - 4)$

❷ 2 terms · 2 terms = 4 partial products

❸ $(x^2 - 3x + 2x - 6)(x^2 - 4)$

❹ $= (x^2 - x - 6)(x^2 - 4)$

❺ 3 terms · 2 terms = 6 partial products

❻ $x^4 - 4x^2 - x^3 + 4x - 6x^2 + 24$

❼ $= x^4 - 10x^2 - x^3 + 4x + 24$

❽ $= x^4 - x^3 - 10x^2 + 4x + 24$

❶ Place square brackets around the first two terms to be multiplied.

❷ Predict the number of partial products for the first two terms.

❸ Multiply each term in the first binomial by each term in the second, producing four products in all (FOIL).

❹ Combine like terms.

❺ Predict the number of partial products.

❻ Multiply each term in the first expression by each term in the second expression.

❼ Combine like terms.

❽ Write the terms of the product in descending order.

Example 2

Expand $x^2(2x^3 + 3xy)(x^3 - 2x^2 - 4)$.

Solution

❶ $[x^2(2x^3 + 3xy)](x^3 - 2x^2 - 4)$

❷ 1 term · 2 terms = 2 partial products

❸ $(2x^5 + 3x^3y)(x^3 - 2x^2 - 4)$

❹ $= 2x^8 - 4x^7 - 8x^5 + 3x^6y - 6x^5y - 12x^3y$

❺ $= 2x^8 - 4x^7 + 3x^6y - 6x^5y - 8x^5 - 12x^3y$

❶ Place square brackets around the first two terms to be multiplied.

❷ Predict the number of partial products for the bracketed terms.

❸ Multiply x^2 by each term in $2x^3 + 3xy$.

❹ Multiply each term in $2x^5 + 3x^3y$ by each term in $x^3 - 2x^2 - 4$. There should be 2 · 3 or 6 terms in all.

❺ You cannot combine like terms since there are none. Arrange the terms in descending order.

Exercises

1. Create a multiplication expression that illustrates the product of a binomial and trinomial. Expand to find the product.

2. Define each term. Give examples from the problem you created in Problem 1.
   (a) term
   (b) factor

   (c) partial product
   (d) distributive property

3. Tell how many partial products will be formed. Then expand and simplify.
   (a) $(x+3)(x^2+4x-3)$
   (b) $(y^2-3y-8)(y+5)$

   (c) $(2+p)(-1-5p+p^2)$
   (d) $\left(x+\dfrac{1}{4}\right)\left(x^2+x-\dfrac{1}{4}\right)$

   (e) $(b^2-4ac)(b^2+4ac-2)$
   (f) $(a+b+c)(a-b)$

4. Expand and simplify.
   (a) $(x-2)(x-2)(x-2)$
   (b) $(x-4)^3$

   (c) $(x-2)(x+3)(5+x)$
   (d) $(3x+5)^3$

   (e) $(-4+x^2)^3$

5. Tell how many partial products will be formed. Then expand and simplify.
   (a) $(x^2+x+1)(x^2+x+2)$

   (b) $(x^2+x+1)^2$

   (c) $(2x^2+2x+2)^2$

6. Expand and simplify.

   (a) $(x^2 + x + 1)(x^2 + x + 2)(2x^2 + 2x + 2)$

   (b) $(x^2 + x + 1)^3$

7. Expand and simplify.

   (a) $8(x + 2)(x + 1)$ **(b)** $-2(a - 5)(a + 6)$

   (c) $(x + 2)(x + 6)4$ **(d)** $-(2m - 3)(m - 3)$

   (e) $2(2x - y)^2$ **(f)** $3(3x + 1)^2$

   (g) $4(x + 3)(x^2 + 4x - 3)$ **(h)** $-2\ (a + b + c)(a - b)$

   (i) $4(y^2 - 3y - 8)(y + 5)$ **(j)** $-x\ (x - 1)(x^2 - x - 1)$

8. Expand and simplify.

   (a) $4(x^2 - 2x + 3)(x^2 + 4x - 3)$ **(b)** $-3(x^2 - 2x + 4)^2$

9. The compound interest formula $A = P(1 + r)^n$ calculates the amount of money (A) that will accumulate over time (n) at a given rate of interest (P). Solve for A, if $n = 3$.

10. A standard die is a cube with six faces. Use the side length shown to find an expression to represent:

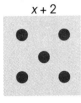

x + 2

(a) the total surface area of the die

(b) the volume of the die

11. A Maya temple in Guatemala has a trapezoidal front face. Use the formula $A = h\left(\dfrac{a+b}{2}\right)$ to show how to calculate the surface area of the exposed face.

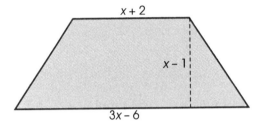

x + 2

x − 1

3x − 6

12. Large stone spheres have been excavated from Maya ruins. Write an expression you can use to calculate the volume of a stone sphere using the formula $V = \dfrac{4}{3}\pi r^3$ if the radius is equal to $(3x^2 + x)$.

13. To expand $5(2x - 7)(x + 4)$, you can multiply the terms in any order to create the partial products. Why?

Use the words *terms* and *commutative property* in your explanation.

4.2 The Greatest Common Factor and Factoring by Grouping

You have learned how to express natural numbers in prime-factored form to find the greatest common factor (GCF).

- To find the GCF, list all the prime factors of each term. The GCF is the product of the factors common to all terms.
- Factor the GCF out of each term. If the first term is negative, factor out the negative of the GCF.

You have also seen how to factor by grouping.
- Rearrange the terms in a different order to facilitate factoring by grouping.
- Group the terms. Factor out the GCF of each group to create a binomial GCF of the expression.
- Factor out the binomial GCF.

To solve a formula for a particular variable:
- Collect all the terms containing that variable on one side.
- Use factoring techniques to isolate the variable.

Example 1

Find the GCF of 36, 48, and 64.

Solution

Express each number in prime-factored form. Then find the factors that are common to all three numbers.

$$36 = \boxed{2} \cdot \boxed{2} \cdot 3 \cdot 3$$
$$48 = \boxed{2} \cdot \boxed{2} \cdot 2 \cdot 2 \cdot 3$$
$$64 = \boxed{2} \cdot \boxed{2} \cdot 2 \cdot 2 \cdot 2 \cdot 2$$
$$\text{GCF} = 2 \cdot 2 = 4$$

Example 2

Factor completely.
$$6x^3y - 8xy^2 + 6x^2y - 8x^2y^2$$

Solution

❶ $2xy(3x^2 - 4y + 3x - 4xy)$
❷ $2xy(3x^2 + 3x - 4xy - 4y)$
❸ $2xy\,[3x(x+1) - 4y(x+1)]$
❹ $2xy[(x+1)(3x-4y)]$
❺ $2xy(x+1)(3x-4y)$

❶ Factor out the GCF.
❷ Rearrange terms in a different order to facilitate factoring by grouping.
❸ Factor out the GCF of each group.
❹ Factor the binomial GCF out of the two terms inside the brackets.
❺ Eliminate redundant brackets.

Example 3

Solve the formula for d, and simplify.

$$bcd + bc^2 = bc - cd$$

Solution

❶ $c(bd + bc) = c(b - d)$
❷ $bd + bc = b - d$
❸ $bd + d = b - bc$
❹ $d(b+1) = b(1-c)$
❺ $d = \dfrac{b(1-c)}{b+1}$

❶ Factor out the GCF, c.
❷ Divide both sides by c.
❸ To isolate terms with d, first subtract bc from both sides and then add d.
❹ Factor d out of the left side and b out of the right side.
❺ Divide both sides by $(b+1)$ to isolate d.

Exercises

1. When we write 100 as $2 \cdot 2 \cdot 5 \cdot 5$, we say that we have written 100 in _____ form.

2. If terms in a polynomial have a GCF of 1, the polynomial is called a _____ polynomial or an irreducible polynomial.

3. Explain why each factorization of $30t^2 - 20t^3$ is not complete.
  (a) $5t^2(6 - 4t)$

  (b) $10t(3t - 2t^2)$

4. Explain why factoring is sometimes described as the reverse of the distributive property. Use $8a^3b^4 - 12a^4b^5$ as an example in your explanation.

5. Find the prime-factored form of each number.
  (a) 98   **b)** 325   **(c)** 288

6. Find the GCF of each set of monomials.
  (a) 16, 40, 60   **(b)** $6x^2y^2z, 9x^3yz^2$ **(c)** $15mnp, 5m^3p, 10mn^2$

7. Factor each polynomial, if possible.
  (a) $2x + 8$ **(b)** $2x^2 - 6x$ **(c)** $3y^3 + 3y^2$

  (d) $5xy + 12ab^2$ **(e)** $13ab^2c^3 - 26a^3b^2c$ **(f)** $14r^2s^3 + 15t^6$

  (g) $25t^6 - 10t^3 + 5t^2$ **(h)** $45x^{10}y^3 - 63x^7y^7 + 81x^{10}y^{10}$

(i) $48u^6v^6 - 16u^4v^4 - 3u^6v^3$

8. Factor out the negative of the greatest common factor for each polynomial.
   (a) $-3a - 6$

   (b) $-6b + 12$

   (c) $-18a^2b - 12ab^2$

   (d) $-21t^5 + 28t^3$

   (e) $-63u^3v^6z^9 + 28u^2v^7z^2 - 21u^3v^3z^4$

9. Factor each expression.
   (a) $4(x + y) + t(x + y)$

   (b) $(a - b)r - (a - b)s$

   (c) $(m + n)^2 + (m + n)$

   (d) $-bx(a - b) - cx(a - b)$

10. Factor by grouping.
   (a) $ar - br + as - bs$

   (b) $x^2 + 4y - xy - 4x$

   (c) $ax + bx - a - b$

   (d) $x^2 + xy + xz + xy + y^2 + zy$

11. Factor by grouping. Factor out all common monomials first.

    (a) $4bc + 4bd - 2bcd - 2bd^2$

    (b) $x^3 - 2x^2y + x^2z - x^2y + 2xy^2 - xyz$

12. Solve for the indicated variable.

    (a) $A = 2wh + 2wl + 6lh$, for h

    (b) $m_1 m_2 = mm_2 + mm_1$, for m_2

13. A four-term polynomial that is factored by grouping can yield two binomial factors, and a six-term polynomial that is factored by grouping can yield a binomial factor and a trinomial factor. What type of factors could an eight-term, a nine-term, and a ten-term polynomial yield? Explain your reasoning and test your hypothesis by working backwards and expanding the factors. Can you devise a rule for predicting the types of factors depending on the number of terms that are factored by grouping?

4.3 Factoring Trinomials and Difference of Squares

In this lesson, you learned these steps for **factoring trinomials** of the form $ax^2 + bx + c$.

- **Find** all possible **pairs of first terms** with a product of ax^2.
- Systematically **combine** the first term pairs with all the possible pairs of factors for c.
- **Look for the combination** that results in the correct value for b. (This is the sum of the products of the outside and inside terms of the binomial factors.)
- **Check the result** by multiplying the binomial factors to see if the product is the original trinomial.

You have also learned how to **factor special types of polynomials**, including:

- perfect square trinomials
- binomials that represent a difference of squares
- trinomials that can be restated in the form $ax^2 + bx + c$ by substituting a variable for part of an expression

Example

Factor $3x^2 - 7x - 6$.

Solution

Step 1

Find all possible pairs of first terms with a product of ax^2.

In $3x^2 - 7x - 6$, ax^2 is $3x^2$, there is one possible combination.

$$(3x \quad)(x \quad)$$

Step 2

Systematically combine the first term pairs with all the possible pairs of factors for c.

In $3x^2 - 7x - 6$, c is -6, so the factors must have a product of -6.

$$(3x + 1)(x - 6)$$
$$(3x - 1)(x + 6)$$
$$(3x + 6)(x - 1)$$
$$(3x - 6)(x + 1)$$
$$(3x + 2)(x - 3)$$
$$(3x - 2)(x + 3)$$
$$(3x + 3)(x - 2)$$
$$(3x - 3)(x + 2)$$

Step 3

Look for the combination that results in the correct value for b.

(This is the sum of the products of the outside and inside terms of the binomial factors.)
In this trinomial, b is -7.

Factor Pair	Numerical Coefficient of x
$(3x + 1)(x - 6)$	$3(-6) + 1(1) = -17$
$(3x - 1)(x + 6)$	$3(6) - 1(1) = 17$
$(3x + 6)(x - 1)$	$3(-1) + 6(1) = 3$
$(3x - 6)(x + 1)$	$3(1) - 6(1) = -3$
$(3x + 2)(x - 3)$	$3(-3) + 2(1) = -7$ ✔
$(3x - 2)(x + 3)$	$3(3) - 2(1) = 7$
$(3x + 3)(x - 2)$	$3(-2) + 3(1) = -3$
$(3x - 3)(x + 2)$	$3(2) - 3(1) = 3$

The only combination that gives -7 as a value for b is $(3x + 2)(x - 3)$.

Step 4

Check the result by multiplying the binomial factors to see if the product is the original trinomial.

$$(3x + 2)(x - 3) = 3x^2 - 9x + 2x - 6$$
$$= 3x^2 - 7x - 6$$

The result is the original trinomial, so the expression has been factored correctly.

Exercises

1. Define each term.
   (a) factor
   (b) polynomial

   (c) trinomial
   (d) binomial

   (e) term
   (f) numerical coefficient

   (g) perfect square trinomial
   (h) difference of squares

2. Explain how you know each set of factors could *not* have the given product. Then give the correct factors for each trinomial.

Factors	Product
(a) $(2p-3)(2p+2)$	$2p^2 - p - 6$

   (b) $(3r+5)(r+5)$ $3r^2 - 20r + 25$

   (c) $(3s+2)(s+1)$ $3s^2 + 7s + 2$

   (d) $(2p-3)(2p-3)$ $4p^2 - 9$

3. Factor.
   (a) $4m^2 + 4m - 15$
   (b) $9n^2 - 30n + 25$

   (c) $25v^2 - 40v + 16$
   (d) $36a^2 - 121$

   (e) $9c^2 - 49d^2$
   (f) $2(h-1)^2 + (h-1) - 21$

   (g) $6(k-2)^2 + (k-2) - 2$
   (h) $2x^4 + 14x^2 + 20$

4. Find all possible integer values of b for which $5x^2 + bx - 1$ can be factored.

5. Find all possible integer values of k for which $7p^2 + kp + 4$ can be factored.

6. A rectangle has an area of $2x^2 + 5x - 12$ square meters. Find expressions for its length and width.

7. The volume of a rectangular box is $3x^3 + 38x^2 + 55x$ in.^{3}. If the box is x in. high, find expressions for its length and width.

8. Greg and Mira have been hired to mow the rectangular athletic field at their school. The field measures 80 yd by 60 yd. Greg will begin at an outside corner and mow around the perimeter in successive rounds until half the field is finished. Then Mira will complete the job. Write an expression to describe the area of the field that Mira will have to mow. Then factor the expression.

9. Is each polynomial a perfect square trinomial, a difference of squares, or neither? Use factors to support your answers.
(a) $9p^2 - 4$

(b) $4x^2 - 6x + 9$

(c) $25r^2 + 10r + 1$

(d) $16 + m^2$

10. Is each statement true or false? Explain.
(a) The last term of a perfect square trinomial is always positive.

(b) The middle term of a perfect square trinomial is always negative.

11. Create a polynomial of each type.

    (a) a perfect square trinomial

    (b) a trinomial that represents a difference of squares

    (c) an $ax^2 + bx + c$ trinomial that can be factored

    (d) an $ax^2 + bx + c$ trinomial that cannot be factored

12. If you are not limited to the integers, can you factor $p^2 - 2$? How?

13. If you are not limited to the integers, can you factor $4x^2 + 2x + \dfrac{1}{4}$? How?

14. Factor $(9x^2 - 6x + 1) - (25y^2 + 40y + 16)$ in as few steps as possible. Compare your work with a classmate's.

4.4 Sum and Difference of Two Cubes

In this lesson, you learned how to factor sums and differences of two cubes.

- Begin by writing each term as the cube of an expression. The base of the first power is F, and the base of the second power L.
- Use these formulas to factor:
  Sum of two cubes: $F^3 + L^3 = (F + L)(F^2 - FL + L^2)$
  Difference of two cubes: $F^3 - L^3 = (F - L)(F^2 + FL + L^2)$

Example 1
Factor: $x^3 + y^3$

Solution

x is the cube root of the first term and y is the cube root of the last term. Use the formula for factoring a sum of two cubes.

$F^3 + L^3 = (F + L)(F^2 - FL + L^2)$
$x^3 + y^3 = (x + y)(x^2 - xy + y^2)$

Example 2
Factor: $b^3 - 64$

Solution

❶ $b^3 - 64 = b^3 - 4^3$
❷ $\quad\quad = (b - 4)(b^2 + b \cdot 4 + 4^2)$
❸ $\quad\quad = (b - 4)(b^2 + 4b + 16)$

❶ Rewrite 64 as a cube. b is the cube root of the first term and 4 is the cube root of the second term.
❷ Use the formula for factoring a difference of two cubes.
❸ Simplify.

Example 3
Factor: $3x^3 + 3$

Solution

❶ $3x^3 + 3 = 3(x^3 + 1)$
❷ $\quad\quad = 3(x^3 + 1^3)$
❸ $\quad\quad = 3(x + 1)(x^2 - x \cdot 1 + 1^2)$
❹ $\quad\quad = 3(x + 1)(x^2 - x + 1)$

❶ Factor out the greatest common factor, 3.
❷ Rewrite 1 as a cube. x is the cube root of the first term and 1 is the cube root of the second term.
❸ Use the formula for factoring a sum of two cubes.
❹ Simplify.

Example 4
Factor: $64 - x^6$

Solution

The expression $64 - x^6$ is a difference of squares with two cubes.

❶ $64 - x^6 = 8^2 - (x^3)^2$
❷ $\quad\quad = (8 - x^3)(8 + x^3)$
❸ $\quad\quad = (2^3 - x^3)(2^3 + x^3)$
❹ $\quad\quad = (2 - x)(2^2 + 2x + x^2)(2 + x)(2^2 - 2x + x^2)$
❺ $\quad\quad = (2 - x)(4 + 2x + x^2)(2 + x)(4 - 2x + x^2)$

❶ Rewrite the expression as a difference of two squares.
❷ Factor the difference of squares. The factors are a difference of cubes and a sum of cubes.
❸ Rewrite the first term in each factor as a cube.
❹ Use the appropriate formula for each factor.
❺ Simplify.

Exercises

1. Write the first ten perfect cubes.

2. Define each term and give an example.

    (a) factor   **(b)** expand

    (c) polynomial   **(d)** binomial

    (e) trinomial   **(f)** term

    (g) difference of two cubes   **(h)** sum of two cubes

3. List similarities and differences between factoring a difference of squares and factoring a sum or difference of cubes.

4. Explain why each factorization is not complete.

    (a) $8x^3 + 64 = (2x + 4)(4x^2 - 8x + 16)$

    (b) $a^6 - 729 = (a^3)^2 - 27^2$
                      $= (a^3 - 27)(a^3 + 27)$

5. Factor completely.

    (a) $x^3 + y^3$ **(b)** $x^3 - 8$ **(c)** $27x^3 + 64y^3$

(d) $27a^3 - b^3$ **(e)** $x^3 + 27$ **(f)** $t^3 - 64$

(g) $64x^3 + 125y^3$

6. Factor out the greatest common factor, then factor the sum or difference of cubes.

    (a) $3x^3 + 81$ **(b)** $2x + 16x^4$

    (c) $4x^3 - 108$ **(d)** $32x - 4x^4$

7. Factor completely.

    (a) $128 - 2x^6$ **(b)** $a^6 - b^6$ **(c)** $8x^6 - 125y^3$

    (d) $125x^3y^6 - 216z^9$ **(e)** $729x^6 - y^6$ **(f)** $64x^6 + 125y^3$

    (g) $a^6 - 729b^6$ **(h)** $8x^3 - 125y^6$ **(i)** $48ax^6 - 6ay^3$

    (j) $3a^6 - 2187b^6$ **(k)** $x^3 - (a^2y)^3$

8. Factor completely.

    (a) $x^3 + (y + z)^3$ **(b)** $(a - b)a^3 + 27$ **(c)** $(x - y)^3 - 125$

(d) $8(m + n) - (m + n)x^3$ **(e)** $x^5y + x^2y^4 - x^3y^3 - y^6$

9. Factor completely.

   (a) $128x^3 - 2x^6$ **(b)** $27x^3 - 1$ **(c)** $5a^3 - 135b^3$

   (d) $(x - y)a^3 + (x - y)z^3$ **(e)** $(xy^2)^3 + 125a^3b^6$ **(f)** $64x^3 + 1$

10. The volume of a box is $(2x^3 + 54)$ in.^{3}. Factor to find the dimensions of the box in terms of x.

11. To make the Italian dessert chocolate-raspberry *tartufo*, a small ball of raspberry ice cream is wrapped in a layer of chocolate ice cream.

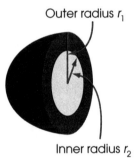

Outer radius r_1

Inner radius r_2

   (a) Write the formula for finding the volume of chocolate ice cream, if the volume of a sphere is $V = \frac{4}{3}\pi r^3$.

   (b) Factor your formula from (a).

   (c) Find the volume of chocolate ice cream if $r_1 = 3.5$ in. and $r_2 = 1.5$ in.

12. Create an expression that is the sum of two cubes and have a partner factor it.

13. Create an expression that is the difference of two cubes and have a partner factor it.

4.5 Dividing Polynomials by Binomials

You have learned how to **divide a polynomial by a binomial** by applying the same **four steps** you use during long division with integers:
- Divide
- Multiply
- Subtract
- Bring down

Repeat these steps until you get a **numerical remainder** or until the **degree** of the remainder, or exponent of the variable, is less than the degree of the divisor.

You have also learned to express your answer in two different forms and to verify your answer.

Form 1: $\dfrac{dividend}{divisor} = quotient + \dfrac{remainder}{divisor}$

Form 2: $dividend = (divisor)(quotient) + remainder$

Example 1

Simplify $(2x^3 - 5x^2 + 7x - 10) \div (2x - 3)$.

Express the result as a division statement in this form:

$\dfrac{dividend}{divisor} = quotient + \dfrac{remainder}{divisor}$

Solution

Rewrite the problem in long division format and then follow the four steps: divide, multiply, subtract, and bring down.

$$
\begin{array}{r}
x^2 - x + 2 \\
2x-3 \overline{) 2x^3 - 5x^2 + 7x - 10} \\
\underline{2x^3 - 3x^2} \\
-2x^2 + 7x \\
\underline{-2x^2 + 3x} \\
4x - 10 \\
\underline{4x - 6} \\
-4
\end{array}
$$

So, $\dfrac{2x^3 - 5x^2 + 7x - 10}{2x - 3} = x^2 - x + 2 + \left(\dfrac{-4}{2x-3}\right)$.

Example 2

Simplify $\dfrac{4x^3 + 5x}{2x - 1}$.

Express the result as a division statement in this form:

$dividend = (divisor)(quotient) + remainder$

Solution

Rewrite the problem in long division format, and then follow the four steps: divide, multiply, subtract, and bring down.

Since there is no x^2-term in $4x^3 + 5x$, insert the place holder $+ 0x^2$. Since there is no constant term, insert the place holder $+ 0$.

$$
\begin{array}{r}
2x^2 + x + 3 \\
2x-1 \overline{) 4x^3 + 0x^2 + 5x + 0} \\
\underline{4x^3 - 2x^2} \\
+ 2x^2 + 5x \\
\underline{+ 2x^2 - x} \\
6x + 0 \\
\underline{6x - 3} \\
3
\end{array}
$$

So, $4x^3 + 5x = (2x - 1)(2x^2 + x + 3) + 3$.

Use this statement to verify the quotient:

$(2x - 1)(2x^2 + x + 3) + 3$
$= 4x^3 + 2x^2 + 6x - 2x^2 - x - 3 + 3$
$= 4x^3 + 5x$

Since the result is the same original dividend, the quotient is correct.

Exercises

1. In this statement, identify the *dividend*, the *divisor*, the *quotient*, and the *remainder*.

$$\frac{x^2 + 10}{x - 4} = x + 4 + \frac{26}{x - 4}.$$

2. Explain the meaning of each term from Problem 1.

3. Anji wrote the following statement to show the results of a division problem:

$$3p^3 - 2p^2 + 6p - 8 = (3p^2 + 4p + 14)(p - 2) + 20$$

   (a) What was the division problem?

   (b) Did Anji solve the problem correctly? Check by multiplying.

4. Here is Ramon's solution to the same problem stated in Problem 3.
   (a) Insert the correct dividend and divisor, and then identify Ramon's errors. Rewrite the solution correctly.

$$
\begin{array}{r}
3p^2 - 8p + 10 \\
\hline
\text{Divisor}\,)\,\text{Dividend} \\
3p^3 - 6p^2 \\
\hline
-8p^2 + 6p \\
-8p^2 + 16p \\
\hline
10p - 8 \\
10p - 20 \\
\hline
-28
\end{array}
$$

   (b) Write a paragraph to explain how Ramon could avoid making similar errors in other division problems.

5. Divide $(b^2 + 6b - 18)$ by $(b + 8)$.

6. Simplify $(4m^2 - 6m - 5) \div (m - 4)$.

7. Simplify $\dfrac{3q^3 + 5q^2 + 4q - 3}{3q - 1}$.

8. Simplify $(8x^2 + 16 + 4x^4) \div (-4 + x)$.

9. Simplify $\dfrac{1 + x + 12x^3}{1 + 2x}$.

10. Divide $(m^4 - m^3 + m^2 - m)$ by $(m^2 + 1)$.

11. Find the length of the rectangle.

$A = (6x^2 - 24)$ in.^{2} $(2x + 4)$ in.

12. The area of a rectangle is $(6x^3 - 2x + 18x^2 - 6)$ in.^{2}. If the length is $(3x^2 - 1)$ in., find the perimeter.

13. The population of a bacteria colony after four days is represented by the formula:

$P = P_0(r^4 + 4r^3 + 6r^2 + 4r + 1)$

P_0 represents the initial number of bacteria, and r represents the rate of increase in the number of bacteria per day, expressed as a decimal. If the original number of bacteria can be represented by $P_0 = \dfrac{336}{1+r}$, find a simplified expression that represents the number of bacteria after four days.

14. Evaluate to the nearest whole number the expression you found in Problem 13 if $r = 12\%$, or 0.12.

15. Refer to Problem 13. If the original number of bacteria is $P_0 = \dfrac{460}{1+r}$ and $r = 15\%$, determine the number of bacteria after four days. Round to the nearest whole unit. Hint: You can use the work you did in Problem 14 to help you find a shortcut to solving this problem.

Discuss Problems 16 to 18 in a small group.

16. What does it mean if you get a remainder of zero when you divide one polynomial by another?

17. Is $(2k - 5)$ a factor of $(2k^3 + 10 - 13k^2 + 16k)$?

18. Find the value of c for which $(2x + 3)$ is a factor of $(6x^2 + 5x + c)$.

5 RATIONAL EXPRESSIONS

5.1 Rational Expressions: Finding Equivalent Forms

In this lesson, you have learned to **simplify rational expressions** by factoring the numerator and the denominator and then eliminating common factors.
- You have found common factors, the difference of squares, and factors of trinomials.
- You have learned that you can change the value of a factor by factoring out –1.
- You have seen how simplified rational expressions can be used to solve measurement or business problems in a very efficient way.

Example

Simplify. $\dfrac{(6x^2 + x - 1)(8x - 4)}{2x - 32x^5}$

Solution

Factor the numerator and the denominator. Then eliminate common factors.

$\dfrac{(6x^2 + x - 1)(4)(2x - 1)}{(2x)(1 - 16x^4)}$ Look for common factors in each expression. You can factor out 4 in the numerator and $2x$ in the denominator.

$= \dfrac{(4)(6x^2 + x - 1)(2x - 1)}{(2x)(1 + 4x^2)(1 - 4x^2)}$ Look for a difference of squares. Factor $(1 - 16x^4)$.

$= \dfrac{(4)(6x^2 + x - 1)(2x - 1)}{(2x)(1 + 4x^2)(1 + 2x)(1 - 2x)}$ Factor $(1 - 4x^2)$.

$= \dfrac{(4)(3x - 1)(2x + 1)(2x - 1)}{(2x)(1 + 4x^2)(1 + 2x)(1 - 2x)}$ Look for a factorable trinomial. Factor $(6x^2 + x - 1)$.

$= \dfrac{(4)(3x - 1)(2x - 1)}{(2x)(1 + 4x^2)(1 - 2x)}$ Look for common factors in the numerator and the denominator, re-ordering terms as needed. Eliminate $2x + 1$ and $1 + 2x$.

$= \dfrac{(4)(3x - 1)(-1)(-2x + 1)}{(2x)(1 + 4x^2)(1 - 2x)}$ You can find another common factor if you factor –1 out of $(2x - 1)$ to get $(-1)(-2x + 1)$ or $(-1)(1 - 2x)$.

$= \dfrac{(4)(3x - 1)(-1)}{(2x)(1 + 4x^2)}$ Eliminate $1 - 2x$ in the numerator and the denominator.

$= \dfrac{(2)(2)(3x - 1)(-1)}{(2)(x)(1 + 4x^2)}$ Factor 2 out of 4 (numerator) and $2x$ (denominator).

$= \dfrac{(2)(3x - 1)(-1)}{(x)(1 + 4x^2)}$ Eliminate the common factor 2.

$= \dfrac{(-2)(3x - 1)}{(x)(1 + 4x^2)}$ Multiply $2 \cdot (-1)$ in the numerator. Check to make sure there are no more common factors.

$x \neq 0,\ \pm\dfrac{1}{2}$ Note the restrictions on x, since the value of the denominator cannot be 0.

Exercises

1. How is simplifying a rational expression like simplifying a rational number? How is it different?

2. Simplify.

(a) $\dfrac{28x^2y^3}{21xy}$

(b) $\dfrac{12p^2 - 30p}{6p}$

(c) $\dfrac{45t^4 - 18t^2}{27t^3}$

(d) $\dfrac{8a^2 - 4a}{6a^2 - 4a}$

(e) $\dfrac{12t - 4a}{6t - 2a}$

(f) $\dfrac{25c^2 - 9d^2}{5c + 3d}$

(g) $\dfrac{16a^2 - b^4}{8a - 2b^2}$

(h) $\dfrac{4y^2 - 4y + 1}{8y - 4}$

(i) $\dfrac{2y^2 - 5y - 3}{2y^2 + 9y + 4}$

(j) $\dfrac{x^2 - r^2}{2r^2 + xr - 3x^2}$

(k) $\dfrac{2e^2 - e - 10}{20 - 3e - 2e^2}$

(l) $\dfrac{x^4 - 81}{x^2 - 2x - 3}$

3. Simplify.

(a) $\dfrac{(x^2 + x - 6)(x^2 + 5x + 4)}{(x^2 + 2x - 8)(x^2 - x - 6)}$

(b) $\dfrac{4y^2 - 8y - 12}{27 - 3y^2}$

(c) $\dfrac{(6\pi a^2 + \pi a - \pi)\,(5a - 3)}{20\pi a^2 - 2\pi a - 6\pi}$

(d) $\dfrac{r^4 - 81}{9 - r^2}$

(e) $\dfrac{(2p - 1)^2 (2p + 1)}{(4p^2 - 1)\,(1 - 2p)}$

4. Which statements are not always true? Explain how you know.
   (a) $4x - 3y = 3y - 4x$ **(b)** $7a + 4b = 4b + 7a$

   (c) $2 - 5t = -1\,(5t - 2)$ **(d)** $7 - 4x = -1\,(7 - 4x)$

   (e) $(4 - t)\,(4 - t) = (t - 4)\,(t - 4)$ **(f)** $(3 + a)\,(3 - a) = (a + 3)\,(a - 3)$

5. Four students each solved a problem in different ways.
   (a) Which solution is correct?

   (b) Describe the errors in each of the other solutions.

Solution 1

$$\frac{6x^2+7x-3}{6x^2-11x+3} = \frac{(2x+3)(3x-1)}{(2x+3)(3x-1)}$$
$$= 1$$

Solution 2

$$\frac{6x^2+7x-3}{6x^2-11x+3} = \frac{(2x+3)(3x-1)}{(2x-3)(3x-1)}$$
$$= \frac{2x+3}{2x-3}$$
$$= \frac{5x}{-x}$$
$$= -5x$$

Solution 3

$$\frac{6x^2+7x-3}{6x^2-11x+3} = \frac{(2x+3)(3x-1)}{(2x+3)(3x-1)}$$
$$= \frac{2x+3}{2x-3}$$
$$= \frac{+3}{-3}$$

Solution 4

$$\frac{6x^2+7x-3}{6x^2-11x+3} = \frac{(2x+3)(3x-1)}{(2x-3)(3x-1)}$$
$$= \frac{2x+3}{2x-3}$$

6. An architect has created two different cottage floor plans. Plan A has a length 3 ft greater than the width. Plan B is larger, with the width and length each 3 ft greater than for Plan A.

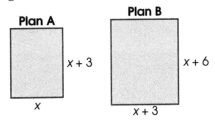

Plan A

Plan B

$x + 3$

$x + 6$

x

$x + 3$

(a) Write a ratio to express the relationship between the floor space of Plan A and the floor space of Plan B. Then simplify the ratio.

(b) If Plan A is 6 ft wide, evaluate the ratio you found in part (a).

7. Create a problem that involves the simplification of a rational expression. Write the solution to your problem on another sheet of paper. Then exchange problems with a classmate and solve.

5.2 Non-permissible Values

In this lesson, you learned that no factor in the denominator of an expression can equal 0, because the value of the expression would be **undefined**. A value that gives this result is called a **non-permissible value**.

You have seen how to determine non-permissible values by inspection, as well as by factoring the denominator and setting each factor to 0. You have also checked non-permissible values by substituting them into expressions.

Example 1

Find the non-permissible values for x.

$$\frac{(x-1)3x}{(x+1)(x-2)}$$

Solution

Since the denominator is already factored, you can identify the non-permissible values for x by inspection.

If $x + 1 = 0$, then $x = -1$.
If $x - 2 = 0$, then $x = 2$.

The non-permissible values for x are -1 and 2.

To check, substitute each non-permissible value into the expression. The non-permissible value is correct if the value of the expression is undefined.

Substitute in $x = -1$.

$$\frac{(x+1)3x}{(x+1)(x-2)} = \frac{(-1+1)(3)(-1)}{(-1+1)(-1-2)} = \frac{(0)(-3)}{(0)(-3)} = \frac{0}{0} \quad \text{The value is undefined.}$$

Substitute in $x = 2$.

$$\frac{(x+1)3x}{(x+1)(x-2)} = \frac{(2+1)(3)(2)}{(2+1)(2-2)} = \frac{(3)(6)}{(3)(0)} = \frac{18}{0} \quad \text{The value is undefined.}$$

Example 2

Find the non-permissible values for a.

$$\frac{4a^2+6}{4a^2-8a-5}$$

Solution

$$\frac{4a^2+6}{(2a+1)(2a-5)} \qquad \text{Factor the trinomial in the denominator.}$$

For the expression to have a denominator equal to 0, either $2a + 1 = 0$ or $2a - 5 = 0$.

$$
\begin{array}{ll}
2a+1=0 & 2a-5=0 \\
2a=-1 & 2a=5 \\
a=-\dfrac{1}{2} & a=\dfrac{5}{2}
\end{array}
$$

The non-permissible values for a are $-\dfrac{1}{2}$ and $\dfrac{5}{2}$.

Exercises

1. Explain why zero is a non-permissible value in the denominator of an expression.

2. Describe a real-life situation where there are non-permissible values.

3. Do the given values make each expression undefined?

(a) $\dfrac{m^2+6}{m^2-9}$, $m=3$

(b) $\dfrac{5y-9}{y^3+y^2-2y}$, $y=2$, $y=-2$

(c) $\dfrac{5mn}{m-2n}$, $m=2n$

(d) $\dfrac{5}{\frac{1}{2}x^2-8}$, $x=-4$, $x=8$

4. What are the non-permissible value(s) for x?

(a) $\dfrac{5}{x}$

(b) $\dfrac{3}{x^3}$

(c) $\dfrac{2x}{x-2}$

(d) $\dfrac{5}{(4x+1)(x-3)}$

(e) $\dfrac{2x+3y}{x^2-y^2}$

(f) $\dfrac{x^2+2x+1}{9x^2-16}$

(g) $\dfrac{5}{x^3-8}$

(h) $\dfrac{2x^2+1}{\frac{1}{2}x^2-8}$

(i) $\dfrac{4x}{0.06x+0.12}$

(j) $\dfrac{4x}{x^2+x-12}$

5. What value for k would make the expression have 1 and -2 as non-permissible values for x?

$$\frac{5x}{4x^2 + kx - 8}$$

6. In the previous lesson, you found this solution to a factoring problem:

$$\frac{6x^2 + 7x - 3}{6x^2 - 11x + 3} = \frac{(2x+3)(3x-1)}{(2x-3)(3x-1)}$$
$$= \frac{(2x+3)}{(2x-3)}$$

(a) What values for x would be non-permissible in the solution?

(b) Does the simplified rational expression have the same non-permissible values as the original expression? Explain.

7. Simplify. Write the non-permissible values for each original expression.

(a) $\dfrac{x^2 - x - 12}{x^2 - 9}$

(b) $\dfrac{2t - 6}{t^2 + 2t - 15}$

(c) $\dfrac{x^2 - 25}{x^2 + 3x - 10}$

(d) $\dfrac{4b^2 - 37b + 9}{b^2 - 81}$

(e) $\dfrac{36 - (x-5)^2}{x^2 + 7x + 6}$

(f) $\dfrac{3x^2 + 2xy - 5y^2}{6x^2 + 7xy - 5y^2}$

8. The formula for finding the area of a rectangle is $A = L \cdot w$, where A is area, L is length, and w is width. The formula can also be expressed as $L = \dfrac{A}{w}$.

   (a) What non-permissible values for w can you find by looking at the formula?

   (b) Use reasoning to suggest other values that are non-permissible for w.

   (c) Are there any restrictions on values for L?

9. The formula relating time to distance and velocity can be written as $t = \dfrac{d}{v}$.

   (a) If you want to travel 500 mi, what will happen to your time as your velocity decreases?

   (b) Explain what happens if the velocity is 0 mph.

   (c) Can the velocity be negative? Explain.

10. Create a rational expression with a denominator that can be factored. Find the non-permissible values for your expression. Then exchange expressions with a partner, find the non-permissible values, and compare results.

5.3 Multiplying and Dividing Rational Expressions

You have seen how multiplying and dividing rational expressions is similar to multiplying and dividing fractions:

- **Factor** each rational expression.
- **Determine any non-permissible values** for variables in the denominators. If you are dividing, include the numerator of the divisor.
- If you are dividing, **restate the problem** by multiplying the dividend by the reciprocal of the divisor.
- **Eliminate common factors** from the numerators and denominators.
- **Multiply** the numerators and the denominators in the simplified expressions.

Example

Divide $\dfrac{14a^2 + 23a + 3}{3 - a - 2a^2} \div \dfrac{7a^2 + 15a + 2}{2a^2 - 3a + 1}$.

Solution

Step 1

Factor.

$$\frac{(2a+3)(7a+1)}{(-2a-3)(a-1)} \div \frac{(7a+1)(a+2)}{(2a-1)(a-1)}$$

Step 2

Determine non-permissible values for a.

Because this is division, include the numerator of the divisor as well as the two denominators.

$$a \neq -\frac{3}{2}, 1, \frac{1}{2}, -\frac{1}{7}, -2$$

Step 3

Multiply the dividend by the reciprocal of the divisor.

$$\frac{(2a+3)(7a+1)}{(-2a-3)(a-1)} \cdot \frac{(2a-1)(a-1)}{(7a+1)(a+2)}$$

Step 4

Restate the problem.

Two factors, $(2a + 3)$ and $(-2a - 3)$, are opposites. Restate $(-2a - 3)$ as $(-1)(2a + 3)$ in the first denominator.

$$\frac{(2a+3)(7a+1)}{(-1)(2a+3)(a-1)} \cdot \frac{(2a-1)(a-1)}{(7a+1)(a+2)}$$

Step 5

Eliminate common factors.

$$\frac{\cancel{(2a+3)}\cancel{(7a+1)}}{(-1)\cancel{(2a+3)}\cancel{(a-1)}} \cdot \frac{(2a-1)\cancel{(a-1)}}{\cancel{(7a+1)}(a+2)}$$

$$= -\frac{1}{1} \cdot \frac{(2a-1)}{(a+2)}$$

Step 6

Multiply the numerators and the denominators.

$$-\frac{2a-1}{a+2}$$

Step 7

Since 1 is non-permissible, check by substituting 2 into each expression.

Original

$$\frac{14a^2 + 23a + 3}{3 - a - 2a^2} \div \frac{7a^2 + 15a + 2}{2a^2 - 3a + 1}$$

$$= \frac{14(4) + 23(2) + 3}{3 - 2 - 2(4)} \div \frac{7(4) + 15(2) + 2}{2(4) - 3(2) + 1}$$

$$= -\frac{105}{7} \div \frac{60}{3}$$

$$= -15 \div 20$$

$$= -\frac{15}{20} = -\frac{3}{4}$$

Simplified

$$-\frac{2a-1}{a+2}$$

$$= -\frac{3}{4}$$

Exercises

1. Define each term and give an example.
   (a) common factor
   (b) reciprocal

   (c) divisor
   (d) dividend

2. Compare multiplication and division of rational expressions to multiplication and division of fractions.

3. Find the non-permissible values and then multiply.

   (a) $\dfrac{7a}{12b^3} \cdot \dfrac{20b^5}{35a^3}$

   (b) $\dfrac{c^2 - 2c - 35}{2c^3 - 3c^2} \cdot \dfrac{4c^3 - 9c}{c - 7}$

   (c) $\dfrac{6 + x - x^2}{x^2 - 13x + 42} \cdot \dfrac{2x^2 - 13x - 7}{2x^2 - 5x - 3}$

   (d) $\dfrac{4m^2 + 15m - 4}{6m^2 + m - 2} \cdot \dfrac{15m^2 + 7m - 2}{5m^2 + 19m - 4}$

4. Find the non-permissible values and then divide.

(a) $\dfrac{12x^6y^2}{55z^3} \div \dfrac{9x^4y^8}{22z^7}$

(b) $\dfrac{x^2+2x-3}{x^2+7x+12} \div \dfrac{x^2+4x-5}{x+1}$

(c) $\dfrac{2y^2-5y-3}{3y^2-10y-8} \div \dfrac{y^2-9}{y^2-y-12}$

(d) $\dfrac{c^2-d^2}{c^2+4cd+3d^2} \div \dfrac{c^2+cd-2d^2}{c^2+cd-6d^2}$

5. Find the non-permissible values and then simplify.

(a) $\dfrac{4x^2}{abc} \div \dfrac{6ac^2}{b^2y^3} \cdot \dfrac{3a^2}{10xy^3}$

(b) $\dfrac{y^2+y}{y^2-4y} \cdot \dfrac{y^2-4y-21}{y+3} \div \dfrac{y^2-6y-7}{y^2-y-12}$

(c) $\dfrac{x^2+3ax}{x-3a} \div \dfrac{x+3a}{x+a} \cdot \dfrac{x^2-4ax+3a^2}{x^2-a^2}$

(d) $\dfrac{9x^2+6x-8}{6x^2+5x-4} \cdot \dfrac{2x^2-7x-4}{2x^2-5x-12} \cdot \dfrac{4x^2+4x-3}{6x^2-x-2}$

(e) $\dfrac{2n^2-7n-15}{5+24n-5n^2} \div \dfrac{2n^2+11n+12}{20n^2+14n+2} \div \dfrac{10n^2+35n+15}{6-7n-3n^2}$

6. Find the non-permissible values and then simplify.

(a) $\dfrac{\dfrac{6a^5b^3}{5xy}}{\dfrac{4a^3b^4}{10xy}}$

(b) $\dfrac{\dfrac{x^2-4}{4x^2}}{\dfrac{x+2}{8x}}$

(c) $\dfrac{\dfrac{2x-4}{x^2+9x+20}}{\dfrac{x^2+x-6}{x^2+7+12}}$

(d) $\dfrac{\dfrac{14a^2+23a+3}{2a^2+a-3}}{\dfrac{7a^2+15a+2}{2a^2-3a+1}}$

7. Caitlin did not receive full marks for this solution. Explain why.

$$\frac{3s^2-14s+8}{2s^2-3s-20} \div \frac{-9s^2+30s-16}{15-34s-16s^2}$$

$$= \frac{(3s-2)(s-4)}{(2s+5)(s-4)} \div \frac{(-3s+2)(3s-8)}{(-8s+3)(2s+5)}$$

$$= \frac{(3s-2)\cancel{(s-4)}}{\cancel{(2s+5)}\cancel{(s-4)}} \cdot \frac{(-8s+3)\cancel{(2s+5)}}{(-3s+2)(3s-8)}$$

$$= \frac{(3s-2)}{1} \cdot \frac{(-8s+3)}{(-3s+2)(3s-8)}$$

$$= \frac{(3s-2)(-8s+3)}{(-3s+2)(3s-8)}$$

8. A circle has a radius of $\dfrac{2x^2+5x-3}{4x^2-12x+5}$.

(a) What expression describes the area of the circle?

(b) What is the area of the circle when $x=3$?

(c) Why can't x be $\dfrac{1}{2}$?

9. A rectangle has an area of $\dfrac{3x+7}{3x+4}$.

(a) If the length is $\dfrac{35+x-6x^2}{3x^2+13x+12}$, what expression describes the width?

(b) What is the width if $x=2$?

(c) Why can't x be greater than $\dfrac{5}{2}$?

5.4 Adding and Subtracting Rational Expressions

You have seen how adding and subtracting rational expressions is **similar to adding and subtracting fractions**:
- Factor each denominator and simplify the expression if necessary.
- Find the lowest common denominator.
- Write equivalent rational expressions using the LCD.
- Add or subtract the expressions.

Example 1

Add $\dfrac{5}{a+7} + \dfrac{4}{a-4}$.

Solution

Step 1

Find the LCD.

The denominators are $(a+7)$ and $(a-4)$.

The LCD is the product of the two denominators, or $(a+7)(a-4)$.

Step 2

Write equivalent rational expressions using the LCD.

$$\frac{5}{a+7} \cdot \frac{a-4}{a-4} = \frac{5(a-4)}{(a+7)(a-4)}$$

$$\frac{4}{a-4} \cdot \frac{a+7}{a+7} = \frac{4(a+7)}{(a+7)(a-4)}$$

Step 3

Add.

$$\frac{5(a-4)}{(a+7)(a-4)} + \frac{4(a+7)}{(a+7)(a-4)}$$

$$= \frac{5a-20}{(a+7)(a-4)} + \frac{4a+28}{(a+7)(a-4)}$$

$$= \frac{5a-20+4a+28}{(a+7)(a-4)}$$

$$= \frac{9a+8}{(a+7)(a-4)}$$

Example 2

Subtract $\dfrac{x-3}{x^2-9} - \dfrac{x+2}{x^2+4x+4}$.

Solution

Step 1

Factor each denominator and simplify the expressions to find the LCD.

$$\frac{x-3}{x^2-9} = \frac{x-3}{(x-3)(x+3)} = \frac{1}{x+3}$$

$$\frac{x+2}{x^2+4x+4} = \frac{x+2}{(x+2)(x+2)} = \frac{1}{x+2}$$

Rewrite the subtraction as $\dfrac{1}{x+3} - \dfrac{1}{x+2}$.

The LCD is the product of the two denominators, or $(x+3)(x+2)$.

Step 2

Write equivalent rational expressions using the LCD.

$$\frac{1}{x+3} = \frac{x+2}{(x+3)(x+2)}$$

$$\frac{1}{x+2} = \frac{x+3}{(x+3)(x+2)}$$

Step 3

Subtract.

$$\frac{x+2}{(x+3)(x+2)} - \frac{x+3}{(x+3)(x+2)}$$

$$= \frac{(x+2)-(x+3)}{(x+3)(x+2)}$$

$$= \frac{x+2-x-3}{(x+3)(x+2)}$$

$$= -\frac{1}{(x+3)(x+2)}$$

Exercises

1. Define each term and give an example.
   (a) lowest common denominator

   (b) rational expression

   (c) non-permissible value

2. Compare addition and subtraction of rational expressions to addition and subtraction of fractions.

3. Identify the lowest common denominator.
   (a) $\dfrac{x}{8}$, $\dfrac{x}{5}$

   (b) $\dfrac{3}{a}$, $\dfrac{5}{a^2}$

   (c) $\dfrac{x}{2y}$, $\dfrac{x+1}{3y}$

   (d) $\dfrac{5}{x-1}$, $\dfrac{7}{x+1}$

   (e) $\dfrac{5}{3(x+3)}$, $\dfrac{2x}{5(x+3)}$

4. Write equivalent rational expressions using the lowest common denominator.
   (a) $\dfrac{2}{x}$, $\dfrac{7x}{5}$, $x \neq 0$

   (b) $\dfrac{x+1}{xy}$, $\dfrac{5}{y^2}$, $x, y \neq 0$

   (c) $\dfrac{x+2}{2(x+1)}$, $\dfrac{x-3}{3(x+1)}$, $x \neq -1$

   (d) $\dfrac{x+3}{x+2}$, $\dfrac{x}{x-2}$, $x \neq \pm 2$

5. Add or subtract. Simplify as necessary.

(a) $\dfrac{3}{a+6}+\dfrac{2}{a-5}$, $a \neq -6, 5$

(b) $\dfrac{5}{a+7}-\dfrac{4}{a-4}$, $a \neq -7, 4$

(c) $\dfrac{t}{2(t+3)}-\dfrac{2}{3(t+3)}$, $t \neq -3$

(d) $\dfrac{6q-1}{81-q^2}-\dfrac{2q}{q+9}$, $q \neq \pm 9$

(e) $\dfrac{4y+16}{(y+4)(y-4)}+\dfrac{4}{y+4}$, $y \neq \pm 4$

(f) $\dfrac{3a}{6a^2+13a+2}+\dfrac{a+1}{a^2+5a+6}$, $a \neq -3, -2, -\dfrac{1}{6}$

(g) $\dfrac{8d}{2d-12}-\dfrac{d+42}{d-6}$, $d \neq 6$

(h) $\dfrac{x+4}{4(x+3)}+\dfrac{3}{6(x-2)}$, $x \neq -3, 2$

(i) $\dfrac{k}{k-1}+k$, $k \neq 1$

(j) $3-\dfrac{x}{x+3}$, $x \neq -3$

(k) $\dfrac{2}{x-1}-\dfrac{x}{2x-1}$, $x \neq -1, \dfrac{1}{2}$

6. Explain why it sometimes helps to simplify rational expressions before you add or subtract. Use an example to illustrate your answer.

7. Write an expression for the perimeter of a triangle with side lengths $\dfrac{2}{x}$, $\dfrac{3}{x+2}$, and $\dfrac{x}{3x-1}$.

8. A boat takes $\dfrac{3}{t+2}$ hours to travel upstream and $\dfrac{5+t}{t+3}$ hours to travel downstream. What expression represents the time the boat takes to complete a round trip?

9. Show how Tabor could have solved this problem more efficiently.

Problem

$$\dfrac{5x+25}{(x+5)(x-2)} - \dfrac{2x-8}{x^2-x-12}$$

Solution

$$\dfrac{2x-8}{x^2-x-12} = \dfrac{2(x-4)}{(x-4)(x+3)}$$

Denominators are: $(x+5)(x-2)$ and $(x-4)(x+3)$

LCD is $(x+5)(x-2)(x-4)(x+3)$

$$\dfrac{5x+25}{(x+5)(x-2)} \cdot \dfrac{(x-4)(x+3)}{(x-4)(x+3)} \qquad\qquad \dfrac{2(x+4)}{(x-4)(x+3)} \cdot \dfrac{(x+5)(x-2)}{(x+5)(x-2)}$$

$$= \dfrac{5x^3+20x^2-85x-300}{(x+5)(x-2)(x-4)(x+3)} \qquad = \dfrac{2x^3-2x^2-44x+80}{(x-4)(x+3)(x+5)(x-2)}$$

Subtract the numerators.

$$\dfrac{5x^3+20x^2-85x-300-(2x^3-2x^2-44x+80)}{3x^3+22x^2-41x-380}$$

The solution is $\dfrac{3x^3+22x^2-44x-380}{(x-4)(x+3)(x+5)(x-2)}$.

10. Create an addition or subtraction problem with rational expressions. Write a solution that contains an error. Then exchange with a classmate to correct one another's solutions.

11. Create an addition or subtraction problem to show that using the LCD can produce an answer that is not in simplest form. Exchange with a classmate to check one another's solutions.

5.5 Solving Rational Equations

In this lesson, you learned these steps for solving rational equations:
- **Factor numerators and denominators** to eliminate any common factors.
- **Identify the lowest common denominator** for the rational expressions.
- **Eliminate the denominators** by multiplying both sides of the equation by the common denominator.
- **Solve the equation** and verify the solution.

Example 1

Solve the equation $\dfrac{1}{2x-3} = \dfrac{1}{9}$, where $x \neq \dfrac{3}{2}$.

Solution

Start by identifying the lowest common denominator.

Since no terms are factorable, the LCD must be the product of the two denominators, $9(2x-3)$.

$$\frac{1}{2x-3} \cdot \frac{9(2x-3)}{1} = \frac{1}{9} \cdot \frac{9(2x-3)}{1}$$ Multiply both sides by the LCD.

$$\frac{9(2x-3)}{2x-3} = \frac{9(2x-3)}{9}$$ Eliminate common factors in each rational expression.

$$9 = 2x - 3$$ Add 3 to both sides of the simplified equation.

$$9 + 3 = 2x - 3 + 3$$

$$12 = 2x$$ Divide both sides of the equation by 2.

$$12 \div 2 = 2x \div 2$$

$$6 = x$$ Verify the solution by substituting 6 for x in the original equation.

L.S.	R.S.
$\dfrac{1}{2x-3}$	$\dfrac{1}{9}$
$= \dfrac{1}{2(6)-3}$	
$= \dfrac{1}{9}$	

Example 2

A tank contains 1,000 L of a solution that is 70% water and 30% salt.
How much water should be boiled off to get a solution that is 35% salt?

Solution

Start by looking for an $\dfrac{a}{b} = c$ relationship. In this case, $\dfrac{\text{volume of salt}}{\text{volume of solution}} = \%$ salt.

If x L of water are to be boiled off, then $1,000 - x$ will be the total volume of the 35% solution.

$$0.35 = \frac{300}{1,000-x}, x \neq 1,000$$ Substitute known values into the equation.

Find any non-permissible values for x.

$$0.35(1,000-x) = 300$$ Multiply both sides by the denominator.

$$350 - 0.35x = 300$$ Simplify the left side.

$$-0.35x = -50$$ Subtract 350 from both sides.

$$x = 142.857$$ Divide both sides by −0.35.

About 143 L of water should be boiled off.

Exercises

1. Explain why multiplying a rational expression by any multiple of the denominator will always eliminate the denominator.

2. Why must any operation that changes one side of an equation be applied to the other side as well?

3. Write an equation that has the operation as a part of its solution. Show the solution.

   (a) An amount is added to both sides.

   (b) Both sides are multiplied by the same amount.

   (c) An amount is subtracted from both sides and then both sides are divided by an amount.

4. Explain how to verify the solution to an equation.

5. Write the non-permissible values for each variable in Problems 6, 7, and 8.

6. Solve and verify. Check to make sure each result is a permissible value for x.

   (a) $\dfrac{x}{5} - \dfrac{3}{8} = \dfrac{17}{40}$

   (b) $\dfrac{5}{2x} - \dfrac{4}{x} = 3$

   (c) $\dfrac{3}{4} = \dfrac{2}{x}$

   (d) $\dfrac{5}{x} = \dfrac{1}{3}$

7. Solve and verify. Check to make sure each result is a permissible value for the variable.

(a) $\dfrac{1}{x+1} = \dfrac{1}{6}$

(b) $\dfrac{1}{3-y} = -\dfrac{1}{10}$

(c) $\dfrac{1}{2x-3} = \dfrac{1}{9}$

(d) $\dfrac{5}{n-25} = \dfrac{3}{n-7}$

8. Solve and verify. Check to make sure each result is a permissible value for the variable.

(a) $\dfrac{1}{x+1} + \dfrac{1}{x-1} = \dfrac{2}{x^2-1}$

(b) $\dfrac{10}{y+3} + \dfrac{15}{y-3} = \dfrac{40}{y^2-9}$

(c) $\dfrac{3}{t-5} - \dfrac{4}{t-5} = \dfrac{2t+4}{t^2-5t}$

(d) $\dfrac{2}{x^2-x} + \dfrac{3}{x^2-1} = \dfrac{4}{x(x-1)}$

9. A wallpaper company uses this formula to find the average cost per roll (in dollars) of producing x rolls of wallpaper in a new pattern.

$$C = \dfrac{15{,}000 + 3.5x}{x}$$

If the average cost is $4.25 per roll, how many rolls is the company planning to produce?

10. A relationship can be expressed by two equivalent ratios, $\dfrac{5}{8+n}$ and $\dfrac{1}{3}$. What is n?

11. Find and correct any errors in Jamilla's solution.

Problem

$$\frac{3}{x+1} - 2 = \frac{5}{3x+3}$$

Solution

First, simplify the denominator $3x + 3$ to $3(x + 1)$.

The LCD must have $(x + 1)$ and $3(x + 1)$ as factors, so you can use $3(x + 1)$.

Eliminate the denominator on the left side by multiplying by the LCD.

$$\frac{3}{x+1} \cdot \frac{3(x+1)}{1} = \frac{9(x+1)}{(x+1)}$$
$$= 9$$

Now eliminate the denominator on the right side.

$$\frac{5}{3(x+1)} \cdot \frac{3(x+1)}{1} = \frac{15(x+1)}{3(x+1)}$$
$$= 5$$

Restate the equation with the simplified expressions.

$\frac{3}{x+1} - 2 = \frac{5}{3x+3}$ becomes $9 - 2 = 5$.

Since $9 - 2 \neq 5$, the equation has no solution.

12. Ashley and Ben each traveled a distance of 3,000 ft. Ashley ran 300 ft/min faster than Ben walked, and she took half as long to run the distance. How fast was each person traveling?

13. If you form a right triangle by dividing an equilateral triangle in half, the hypotenuse of the right triangle is always equal to double the length of the shorter side.

What does x represent in the base length of this triangle?

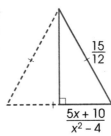

14. Two triangles with areas 15 cm^2 and 24 cm^2 have the same base length. The height of the smaller triangle is 3 cm less than the height of the larger one. What is the common base length?

6 RELATIONS AND FUNCTIONS

Differences Between Relations and Functions

In this lesson, you learned that a **relation** is a property that connects elements from two or more sets. If there are two sets of elements, the relation is called a **binary relation**. A binary relation can be represented by a set of **ordered pairs**.

There are two different types of relations:

Function: A function is a relation where each element in the first set maps to only one element in the second set.

A graph represents a function if no vertical line passes through more than one point on the graph.

A function can be represented by a
• one-to-one mapping.
• many-to-one mapping.

Non-function: In a non-function, at least one element in the first set can be mapped to more than one element in the second set.

A graph represents a relation that is not a function if at least one vertical line passes through more than one point in the graph.

A relation that is not a function can be represented by a one-to-many mapping.

Example

Is the relation defined by $y^2 = x$ a function?

Solution

Method 1: Guess and Test

Choose values of x and determine the corresponding values of y. Look for patterns in the values that will help you predict whether the relation is a function.

Let $x = 4$.

$$y^2 = 4$$
$$y = \pm\sqrt{4}$$
$$y = \pm2$$

Since there is more than one y-value associated with the x-value, the relation is not a function.

Method 2: Use the Vertical Line Test

Graph several values for (x, y) where $y^2 = x$.

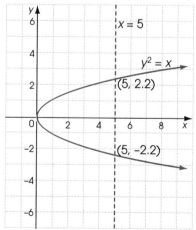

Since at least one vertical line passes through more than one point on the graph, the relation is not a function.

Exercises

1. Identify a relation that exists between the elements of two sets of people you know. Identify the input values and the output values. Explain how you know that the relation is or is not a function.

2. In your own words, define one-to-one mapping, many-to-one mapping, and one-to-many mapping. How are these types of mapping the same? How are they different?

3. A relation is represented by this mapping diagram.

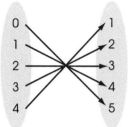

(a) List the ordered pairs of the relation.

(b) Is the relation a function? Explain.

4. Draw a mapping diagram for each set of ordered pairs.
(a) {(0, 0), (1, 1), (2, 2), (3, 3)}

(b) {(0, 0), (1, 0), (2, 2), (3, 2)}

(c) {(0, 0), (1, 1), (2, 3), (2, 4)}

5. Compare the mapping diagrams you drew for Problem 4. How are they the same? How are they different?

6. Which sets of ordered pairs from Problem 4 represent functions? Explain how you know.

7. Draw a mapping diagram for each relation.
   (a) $y = 2x - 3$, $x \in \{-4, -2, 0, 2, 4\}$

   (b) $y = 4x^2$, $x \in \{-3, -1, 0, 1, 3\}$

8. Which relations from Problem 7 are functions? Why?

9. Which ordered pair satisfies each given relation?
   (a) $3x - y = 8$; $(2, 2)$ or $(3, 1)$ **(b)** $y < 2x - 1$; $(2, 3)$ or $(3, 4)$

   (c) $y = x^2 - 2$; $(2, 2)$ or $(-2, -6)$ **(d)** $x^2 + y^2 = 25$; $(3, 4)$ or $(4, 3)$

10. Which graphs represent functions?

   (a)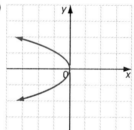

   (b)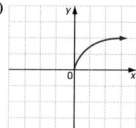

   (c)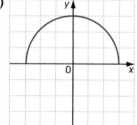

   (d)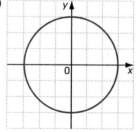

11. A horizontal line is drawn through the graph for a function.

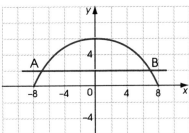

(a) Name the coordinates of A and B.

(b) What type of mapping would represent the function? Why?

12. Identify the type of mapping that would represent each function:

(a)

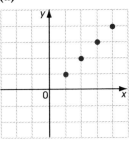

(b)

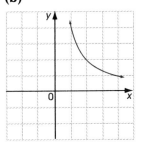

Describing Functions

You have seen that **functions** can be represented in **three ways**:
- as a table of values or set of ordered pairs
- as a graph
- as a rule expressed in word or equation form

Example

Sam proposes a new monthly allowance program to his parents. His first month's allowance will be 1¢. The amount will double in each subsequent month, so Sam will get 2¢ in Month 2, 4¢ in Month 3, and so on. Should Sam's parents accept this arrangement?

Solution

The total allowance Sam's parents will have to pay is a function of the number of months.

Step 1: Create a Table of Values

Number of Months	Allowance Paid This Month (¢)	Total Allowance Paid (¢)
1	1	1
2	2	3
3	4	7
4	8	15
5	16	31
6	32	63

Over the first six months, Sam's parents would have to pay 63¢.

Step 2: Graph the Data

A graph could help Sam's parents predict what would happen over a longer period of time. The number of months is the independent variable, so each ordered pair should be organized as (*number of months, allowance paid in all*).

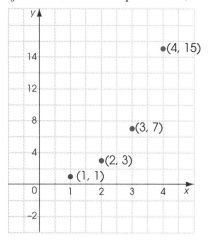

Notice:
- Sam is paid once each month, so the function is discrete. Points are not joined with a curve.
- The sharp upward curve indicates that the function is exponential. The amount of allowance paid rises very sharply as the number of months increases.

Step 3: Find the Equation

To find the exact amount they will have to pay over any number of months, Sam's parents can use an equation.

Look for a number pattern in the table of values that relates the number of months to the amount of allowance paid in all. Remember, the graph indicates that the pattern is exponential.

Number of Months	Allowance Paid in All (¢)
1	$1 = 2^1 - 1$
2	$3 = 2^2 - 1$
3	$7 = 2^3 - 1$
4	$15 = 2^4 - 1$
5	$31 = 2^5 - 1$
6	$63 = 2^6 - 1$

The equation that relates the number of months (n) to the total allowance paid (A) is $A = 2^n - 1$.

Now Sam's parents can determine the total amount they would have to pay after 12 months and after 24 months.

For 12 months: $A = 2^{12} - 1 = 4{,}095¢$
$$= \$40.95$$

For 24 months: $A = 2^{24} - 1 = 16{,}777{,}215¢$
$$= \$167{,}772.15$$

Although this plan sounds good for the first year, it is not feasible for Sam's parents to carry it through the second year.

Exercises

1. Identify the independent variable and the dependent variable.

(a)

Time (h)	Distance Traveled (km)
1	50
2	100

(b)

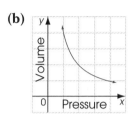

(c) $Area = 12 \cdot length$

2. Explain how to identify the independent variable and the dependent variable of a function when it is represented by:

(a) a table of values or set of ordered pairs

(b) a graph

(c) an equation

3. Complete the ordered pairs for each equation.

(a) $y = \frac{2}{3}x - 5$ $(-1, \underline{\quad}), (\underline{\quad}, 10)$

(b) $y = 3x^2 - 4$ $(-4, \underline{\quad}), (3, \underline{\quad})$

(c) $y = 5 + 2\sqrt{x - 3}$ $(4, \underline{\quad}), (12, \underline{\quad})$

(d) $y = 3^x - 1$ $(2, \underline{\quad}), (4, \underline{\quad})$

(e) $y = x^3 - 2x^2 + 1$ $(-2, \underline{\quad}), (1, \underline{\quad})$

(f) $y = \frac{2}{x+1} - 5$ $(1, \underline{\quad}), (-3, \underline{\quad})$

4. Select the equations that could be satisfied by the ordered pair $(2, 5)$.

(a) $y = 3x - 1$ **(b)** $y = 3 + \sqrt{x + 2}$

(c) $y = x^3 - 3$ **(d)** $y = 2^x + 1$

(e) $y = 2x^2 - 6$ **(f)** $y = 3^x - 1$

5. If $(2, 5)$ is an ordered pair that satisfies $y = -x^2 + k$, then what is the value of k?

6. Determine an equation to represent each table of values.

(a)

x	y
0	1
1	3
2	5
3	7

(b)

x	y
1	24
2	12
3	8
6	4

(c)

x	y
1	-4
2	-1
3	4
4	11
10	95

(d)

x	y
0	0
1	2
4	4
25	10
49	14

7. Match each graph to its equation or rule.

$y = 3^x,\ x \geq 0$ $y = \dfrac{3}{x},\ x \geq 0$

$y = \sqrt{x-2},\ x \geq 2$ $y = \sqrt{x} - 2,\ x \geq 0$

$2 per hour or any portion thereof

$1 per hour or any portion thereof

(a)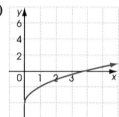

(b)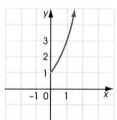

(c)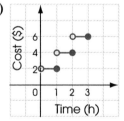

8. In this pattern, the number of dots is a function of the figure number. Represent the function with a set of ordered pairs, a graph, and an equation.

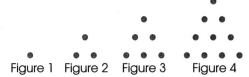

Figure 1   Figure 2   Figure 3   Figure 4

9. How many dots will be in Figure 5 of the pattern above?

6.3 Function Notation

In this lesson, you learned how to use function notation to **evaluate functions** by
- substituting a number for a variable.
- substituting an algebraic expression for a variable.
- locating a point on a graph.

You also learned to evaluate functions that are linked by operations such as addition and multiplication.

Example 1

If $f(x) = \dfrac{1}{x(x+2)}$, find $f(-1)$, $f\left(\dfrac{1}{2}\right)$, and $f(5)$.

Solution

$$f(-1) = \frac{1}{(-1)\,[\,(-1)+2\,]}$$
$$= \frac{1}{(-1)\,(1)}$$
$$= -\frac{1}{1}$$
$$= -1$$

$$f\left(\frac{1}{2}\right) = \frac{1}{\left(\dfrac{1}{2}\right)\left(\dfrac{1}{2}+2\right)}$$
$$= \frac{1}{\left(\dfrac{1}{2}\right)\left(\dfrac{5}{2}\right)}$$
$$= \frac{1}{\dfrac{5}{4}}$$
$$= 1 \div \frac{5}{4}$$
$$= \frac{4}{5}$$

$$f(5) = \frac{1}{(5)\,(5+2)}$$
$$= \frac{1}{5(7)}$$
$$= \frac{1}{35}$$

Example 2

Given the graph of $h(x)$, find $h(1)$, $h(4)$, and $h(-2)$.

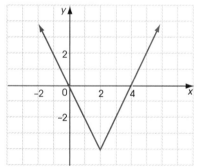

Solution

When x is 1, y is -2, so $h(1) = -2$.

When x is 4, y is 0, so $h(4) = 0$.

When x is -2, y is 4, so $h(-2) = 4$.

Exercises

1. Give two reasons why mathematicians may have introduced function notation.

2. Define the words *input* and *output*.

3. Restate each function using function notation.

(a) $y = \sqrt{x} - 3$ (b) $y = \dfrac{x^3}{x-1}$

4. Explain in words how you would find the output value for a given input value.

(a) $f(x) = -x^2$ (b) $h(n) = \dfrac{n}{4} - n^3$

(c) $g(p) = p(p-2)$ (d) $p(g) = \sqrt{g}$

5. Sketch a graph of the function $g(x)$ such that $g(4) = 7$, $g(-2) = 0$, and $g(0) = -3$.

6. Find each value for $f(x) = \sqrt{x^2 - 2}$. Round your answer to one decimal place where necessary.

(a) $f(6)$ (b) $f(0)$

(c) $f(\sqrt{2})$ (d) $f\left(\dfrac{7}{4}\right)$

(e) $f(x-1),\ x \geq 1$ (f) $f(a^2)$

7. Find each value.

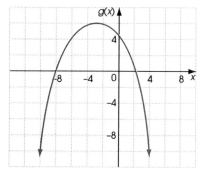

(a) the value of x that gives the greatest value of $g(x)$

(b) the value(s) of x such that $g(x) = 0$

(c) $g(0)$

8. If $p(n) = n^2 - 2$ and $p(n) = 7$, find the value(s) of n.

9. As a rocket takes off and accelerates at a constant rate, the distance traveled is a function of time. This relationship is given by the equation $d(t) = \frac{1}{2}at^2$, where $d(t)$ is the distance traveled in meters, a is a constant acceleration of 25 m/s^2, and t is time in seconds. Find $d(1)$, $d(7)$, and $d(12)$.

10. Complete.
   (a) If $f(x) = 3x - 4$ and $g(x) = \frac{x+4}{3}$, find $f(g(2))$ and $g(f(2))$.

   (b) Repeat with a different input value for x. Is $f(g(x))$ always equal to $g(f(x))$ for these functions?

   (c) If $f(x) = x^2 - 3x$ and $g(x) = 5x + 4$, find $f(g(0))$ and $g(f(0))$.

   (d) Repeat with a different input value for x. Is $f(g(x))$ always equal to $g(f(x))$ for these functions?

11. If $f(a) = a^2 - a - 10$ and $g(a) = 6 - a$, for what value(s) of a is $f(a) = g(a)$?

12. A toy rocket is launched from a park. The height it achieves (in meters) is a function of the time it is in the air (in seconds) and can be represented by the function $h(t) = -0.06t^2 + 6t - 10$.
   (a) Find its height two seconds after it is launched.

   (b) After how many seconds is it at its greatest height?

13. If $f(x) = 2x - 3$ and $g(x) = x^2 - 2$, find each value.
   (a) $f(2) - g(3)$

   (b) $g(-1) - f(-2)$

   (c) $f(-3) + g(0)$

   (d) $f(a + h) - f(a)$

   (e) $f(g(1))$

   (f) $g(f(1))$

   (g) $f(f(2))$

   (h) $g(g(-3))$

   (i) $\dfrac{g(a+h) - g(a)}{h}$

   (j) $\dfrac{f(1)}{g(-1)}$

14. Use the graphs to find each value.

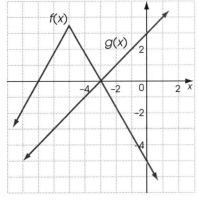

   (a) $f(-4)$

   (b) $g(1)$

   (c) $f(g(-7))$

   (d) $g(f(0))$

   (e) $g(g(0))$

In this lesson, you learned how to determine the **domain** and **range** of a relation by examining its **graph**.
- The domain is the set of *x*-values for a relation.
- The range is the set of *y*-values for a relation.

You also learned how to identify continuous, discontinuous, and discrete data, and examined the impact of each characteristic on the domain and the range of a relation.

Example 1

Determine the domain and range.

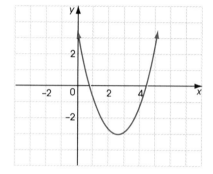

Solution

The graph is continuous because there are no breaks (discontinuities) in the line.

Since the data is continuous, all values for *x* and *y* are real numbers.

The arrows indicate that the curve, a parabola, extends without bounds, so the sets of values for *x* and *y* are infinite.

There are no maximum or minimum values for *x*.

The domain is $\{x \mid x \in R\}$.

The graph has a relative minimum at $(3, -3)$, so all *y*-coordinates are greater than -3.

The range is $\{y \mid y \geq -3, y \in R\}$.

Example 2

Determine the domain and range.

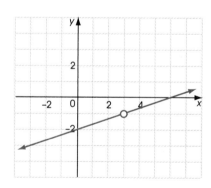

Solution

The arrows indicate that the line extends without bound in each direction. However, the open point at $(3, -1)$ indicates that this point is not on the line.

The domain is $\{x \mid x \neq 3, x \in R\}$. The range is $\{y \mid y \neq -1, y \in R\}$.

Exercises

1. Define the domain of a relation in terms of ordered pairs.

2. Relations are often shown on a graph, where not all of the ordered pairs are known. Give a definition for the domain of a relation in this case.

3. Define the range of a relation in terms of ordered pairs.

4. Define the range of a relation shown on a graph.

5. The domain of a relation is $\{x \mid x \neq 5, x \in I\}$. What can you predict about the graph?

6. The range of a relation is $\{y \mid y \neq 0, y \in R\}$. What can you predict about the graph?

7. The domain of a relation is $\{x \mid x \in R\}$, and the range is $\{y \mid y = 7, y \in R\}$. Describe the graph.

8. The domain of a relation is $\{x \mid x = 5, x \in I\}$, and the range is $\{y \mid y \in I\}$. Describe the graph.

9. Determine the domain and range.
   (a)
   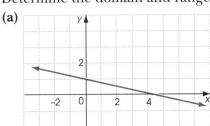

(b)

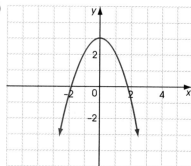

(c)

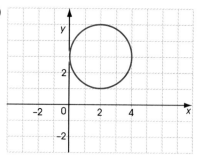

(d)

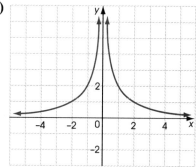

10. Determine the domain and range.

(a)

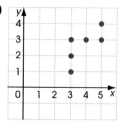

(b)

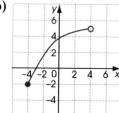

(c)

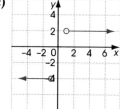

(d)

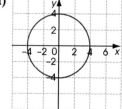

11. Construct the graph of a circle whose domain is $\{x \mid -5 \leq x \leq 1, x \in R\}$.

12. This graph illustrates the hourly cost of renting a post hole auger from an equipment rental company.

Determine the domain and range of the relation, and explain what the numbers mean to a consumer.

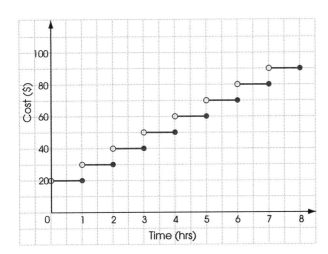

7 LINEAR FUNCTIONS

7.1 Linear Functions

In this lesson, you learned how to use the equation $y = mx + b$ to find the **intercepts**, **slope**, **domain**, and **range** of a **linear function**.

- The slope of a straight line is defined by $\frac{rise}{run}$ or $\frac{y_2 - y_1}{x_2 - x_1}$. In $y = mx + b$, m represents the slope.
- The y-intercept is where a graph meets the y-axis. In $y = mx + b$, b represents the y-intercept.
- The x-intercept is where a graph meets the x-axis. To find the x-intercept, substitute 0 for y in $y = mx + b$, and then solve for x.
- The **domain of a relation** is the set of all x-values, and the **range** is the set of all y-values.

Example 1

A long-distance telephone company charges $8 per month plus 25¢ per minute for calls to Europe. Write an equation for the cost, C (in dollars), in a month when t minutes of calls are made to Europe and no other long-distance calls are made. Find the slope of the line and the C-intercept.

Solution

The cost of one month's telephone use is $8 plus 25¢ times the number of minutes spent calling Europe.

$C = 8 + 0.25t$

Express the fixed and variable costs in dollars.

To find the slope and C-intercept, write the equation in the form $y = mx + b$. Since cost depends on time, C must be the y-value. The independent variable t replaces x in the equation.

$$y = mx + b$$
$$C = 0.25t + 8$$
$$\text{Slope } = 0.25$$

The slope is the coefficient of t.

The C-intercept is represented by b, which is 8 in this equation. The graph meets the y-axis at $y = 8$, so the least possible cost for one month's service is $8.

Example 2

A straight line crosses the y-axis at –1 and the x-axis at 3. What equation does the graph represent?

Solution

To state the equation of a linear relation in the form $y = mx + b$, you need to know the slope (m) and the y-intercept (b).

The line crosses the y-axis at –1. Substitute –1 for b to get $y = mx - 1$.

To find the slope, locate any two points on the line and use $\frac{y_2 - y_1}{x_2 - x_1}$. You already know the coordinates of two points on the line, the y-intercept at (0, 1) and the x-intercept at (3, 0).

$$\frac{y_2 - y_1}{x_2 - x_1} = \frac{0 - 1}{3 - 0} = -\frac{1}{3}$$

Substitute $-\frac{1}{3}$ for m in the equation $y = mx - 1$.

The equation of the line is $y = -\frac{1}{3}x - 1$.

Example 3

The speed of a model car as it slows down is given by the equation $v = 20 - 4t$, where v is the speed in meters per second and t is the time in seconds during which the car has slowed.

Give the slope of the line, the t-intercept, and the domain and range. Explain the meaning of the domain and range.

Solution

Write the equation in the form $y = mx + b$.

$$v = -4t + 20 \qquad \text{Slope} = -4$$

The slope is the coefficient of the independent variable, t.

To find the t-intercept, substitute 0 for the dependent variable, v, and then solve the equation for t.

$$
\begin{aligned}
0 &= -4t + 20 \\
-20 &= -4t \qquad &\text{Subtract 20 from both sides.} \\
5 &= t \qquad &\text{Divide both sides by } -4.
\end{aligned}
$$

The domain is the set of values of t. The time cannot be negative, so $t > 0$.

The speed reaches 0 m/s where the graph intersects the t-axis at $t = 5$, so t cannot be greater than 5.

The change in speed is continuous, so t can be any real number between 0 and 5. The domain represents all possible numbers of seconds between the car's starting and stopping times.

The range is the set of values of v. The speed starts at 20 m/s and reduces to 0 m/s.

The range is all real values of v between 0 and 20. These values represent all possible speeds of the car while it is slowing down.

Exercises

1. Explain why you can find the x-intercept by substituting 0 for y in the equation $y = mx + b$.

2. Explain why the formula $\dfrac{y_2 - y_1}{x_2 - x_1}$ is used to determine the slope for a straight line.

3. State the domain and range of each line.

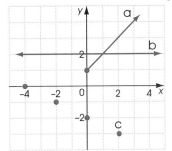

4. State the slope and intercepts of each line.

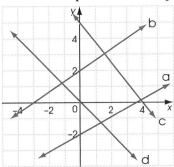

5. Write the equation for each line in Problem 4.

6. Graph each line. Give the x- and y-intercepts.

    (a) $y = 2.5x - 5$ **(b)** $y = \frac{2}{3}x + 2$ **(c)** $y = 4 - 2x$

7. Graph.

    (a) $y = 3 - x$ for $x = -1, 0, 1, 2,$ or 3

    (b) $y = \frac{3}{2}x - 3$ for $x \geq 0, y \geq 0$

    (c) $y = -\frac{1}{2}x - 1$ for $x \neq -2$

8. State the y-intercept and the slope.

    (a)

x	-2	-1	0	1	2	3
y	8	6	4	2	0	-2

    (b)

x	-2	0	2	4	6
y	-9	-6	-3	0	3

9. Give the domain and range.

    (a) $y = 6 - 4x$, where $x = 0, 1, 2, 3, 4$

    (b) $y = \frac{1}{2}x + 1$, where x is a positive integer ≤ 5

    (c) $y = 3x - 5$, where x is a real number

    (d) $y = 2x + 1$, where x is a real number ≥ 0

10. Graph each line on the same coordinate grid. What do you notice?

   (a) $y = 7 - x$ **(b)** $y = 3$ **(c)** $y = 1.5x - 3$

11. State the slope of each line in Problem 10.

12. State the slope and the x- and y-intercepts.

   (a) $y = 5 + 2x$ **(b)** $y = \frac{2}{3}x - 4$

   (c) $y = -\frac{1}{2}x - 3$ **(d)** $y = 6 - 1.5x$

13. Can you sketch both lines? If not, explain.

   (a) The slope is negative and the x- and y-intercepts are positive.

   (b) The slope is positive and the x- and y-intercepts are negative.

14. Which lines have an undefined slope?

   (a) $y = 4$ **(b)** $x = -1$ **(c)** $y = -2x$

   (d) $y = -\frac{1}{2}$ **(e)** $x = 3$ **(f)** $y = x$

15. Which lines have a negative slope and positive x- and y-intercepts?

   (a) $y = 5 - 2x$ **(b)** $y = 3x - 5$

   (c) $y = -\frac{1}{3}x - 2$ **(d)** $3 - x = y$

16. State the domain, range, and slope for each line segment.

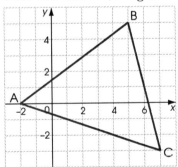

17. Draw a parallelogram on a coordinate grid.

   (a) Give the equations of the sides.
   (b) Give the domain and range for each.

You have seen that a linear relation can show either a **direct** or a **partial variation**.
- If y varies directly with x, the two variables **increase or decrease together** and the graph **passes through the origin**. The equation for a relation of this type is $y = kx$, where k is the **constant of variation** or slope of the line.
- If y varies partially with x, the variables still increase or decrease together, but the y-intercept is not at the origin. The equation for a partial variation is $y = kx + b$ or $y = b + kx$, where b is the y-intercept and k is the slope of the line.

Example

The amount that a spring stretches varies directly with the weight attached to it. If the spring stretches 2.5 in. to hold 5 lb, what weight will stretch the spring 56 in.?

Solution

Method 1: Use the Constant of Variation (k)

Let s represent the amount of stretch in inches.
Let w represent weight in pounds.

The amount of stretch depends on the weight, so w is the x-value and s is the y-value.

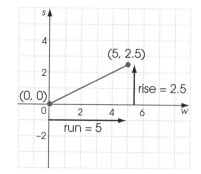

You know the line passes through $(0, 0)$ because the variation is direct. You can use the distance between $(0, 0)$ and $(5, 2.5)$ to determine the slope of the line (k).

$$Slope = \frac{rise}{run}$$
$$k = \frac{s}{w}$$
$$k = \frac{2.5}{5}$$
$$k = 0.5$$

The equation that represents a direct variation is $y = kx + 0$ or $y = kx$.
Rewrite the equation, substituting s for y and w for x.

$s = kw$	Substitute 0.5 for k.
$s = 0.5w$	
$56 = 0.5w$	Solve the equation for $s = 56$.
$\frac{56}{0.5} = \frac{0.5w}{0.5}$	Divide both sides by 0.5.
$112 = w$	

The spring will stretch 56 in. when the weight is 112 lb.

Method 2: Solve a Proportion

Since the variation is direct, the relationship between 2.5 in. and 5 lb is the same as between 56 in. and w lb.

$\frac{2.5}{5} = \frac{56}{w}$	Cross multiply.
$2.5 \cdot w = 5 \cdot 56$	Multiply both sides.
$2.5w = 280$	
$\frac{2.5w}{2.5} = \frac{280}{2.5}$	To isolate w on one side, divide both sides by 2.5.
$w = 112$	

The string will stretch 56 in. when the weight is 112 lb.

Exercises

1. Define each term.
   (a) direct variation   (b) partial variation

   (c) constant of variation   (d) dependent variable

   (e) independent variable   (f) fixed cost

   (g) variable cost

2. Explain the difference between direct variation and partial variation. Give an example of each that includes a set of ordered pairs, a graph, and an equation.

3. Find the constant of variation.
   (a) m varies directly as n, and $m = 28$ when $n = 14$

   (b) p is directly proportional to q, and $p = 36$ when $q = 3$

   (c) $r \propto s$ and $r = 120$ when $s = 24$

4. Complete.
   (a) If x is directly proportional to y, write an equation to express the direct variation.

   (b) If $x = 25$ when $y = 16$, find y when $x = 100$.

5. If $y \propto x$, find the missing values for each direct variation.
   (a) If $x = 4$ when $y = 10$, what is y when $x = 20$?

   (b) If $x = 3.6$ when $y = 1.8$, what is y when $x = 7.2$?

6. A partial variation is given by $B = 18 + 6n$. Find B for each given value of n.
  (a) $n = 12$
  (b) $n = 25$

  (c) $n = 36$
  (d) $n = 0.25$

7. A partial variation is given by $W = 150 + 20p$. Find p for each given value of W.
  (a) $W = 390$
  (b) $W = 790$

  (c) $W = 490$
  (d) $W = 1,330$

8. The cost of a 30 cm pizza with tomato sauce and cheese is $6.00. Each additional topping costs $0.75.
  (a) Write an equation to show the partial variation.

  (b) Construct a table of values for 1 to 8 additional toppings.

  (c) Draw a graph of the partial variation.

9. The mass of a substance varies directly with its volume.
  (a) Write an equation to express the direct variation.

  (b) When the mass is 300 g the volume is 240 cm^3. Find the mass when the volume is 384 cm^3.

10. The cost of parking in a downtown car lot varies directly with the length of time parked. If Gabe paid $10.50 for 6 h, what is the charge for 2 h?

11. Marla is a sales manager. She receives a fixed salary of $3,500 per month plus a variable amount calculated as $\frac{1}{2}$% commission on the total sales made by her staff. Find her total December income if the sales for that month were $625,000.

12. When Jannelle telephones her sister, the cost is $3.50 for the first three minutes plus $0.65 for each additional minute.

   (a) Find the cost for an 8 min phone call from Jannelle to her sister.

   (b) Jannelle's last call to her sister cost $11.30. How long was the call?

13. On a scale model of a sailboat, an object that is actually 2 m long is represented by a similar object 25 cm long. How tall is the mast on the model if the actual mast height is 12 m?

14. The number of words you can type is directly proportional to the time you spend typing. If Jim can type 225 words in 5 min, how long will it take him to type an essay with 1,125 words?

15. Trevor rented a car for one day from an agency that charges a flat rate of $40 per day plus 10¢/mi. Trevor drove the car 326 mi before returning it. When the rental agent tried to calculate the amount to charge, he knew he had made an error because the result was much too high.

   (a) Where did the rental agent make his error?

   (b) What should the correct cost be?

The Agent's Solution
The fixed cost is $40.
The variable cost is 10¢ for each mile.
Let C represent cost and d represent distance.
$C = 40 + 10d$
$C = 40 + 10(326)$
$C = 40 + 3,260$
$C = \$3,300$
Trevor should pay $3,300.

In this lesson, you learned how to **calculate the slope of a line or line segment**. For any two points, $A(x_1, y_1)$ and $B(x_2, y_2)$, the slope, m, of AB can be written in one of the following ways:

$$m = \frac{rise}{run} \qquad m = \frac{y_2 - y_1}{x_2 - x_1}$$

- If $m > 0$, the line segment is oblique and rises to the right.
- If $m = 0$, the line segment is horizontal.
- If $m < 0$, the line segment is oblique and falls to the right.
- If m is undefined, the line segment is vertical.

Example 1

Calculate the slope of the line segment joining $A(-2, -4)$ to $B(3, 5)$.

Solution

Let (x_1, y_1) be $(-2, -4)$ and (x_2, y_2) be $(3, 5)$.

$$m = \frac{y_2 - y_1}{x_2 - x_1}$$

$$m = \frac{5 - (-4)}{3 - (-2)}$$

$$m = \frac{9}{5}$$

The slope of line segment AB is $\frac{9}{5}$.

If line segment AB had been plotted on a coordinate grid, the slope could also have been calculated by using the formula $m = \frac{rise}{run}$.

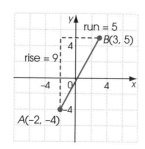

$$m = \frac{rise}{run}$$

$$m = \frac{9}{5}$$

Example 2

If two points on a line are $C(5, 4)$ and $D(8, 6)$, find another point on the line.

Solution

Step 1

Calculate the slope of line segment CD.

Let (x_1, y_1) be $(5, 4)$ and (x_2, y_2) be $(8, 6)$.

$$m = \frac{y_2 - y_1}{x_2 - x_1}$$

$$m = \frac{6 - 4}{8 - 5}$$

$$m = \frac{2}{3}$$

The slope of line segment CD is $\frac{2}{3}$.

Step 2

Plot either point and then use the slope to find another point on the same line.

$$m = \frac{2}{3} = \frac{rise}{run}$$

The rise is 2 units and the run is 3 units. To find the new point, move 2 steps up and 3 steps to the right.

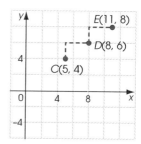

Another point on line segment CD is $E(11, 8)$.

Exercises

1. In your own words, explain what slope is, how it can be calculated, and how it can be predicted.

2. Write sentences that incorporate two or more of these terms. (Use the terms as they relate to slope calculations.) Continue until you have used all the terms at least once.

negative	*flatter*	*slope*
positive	*run*	*vertical*
rise	*steeper*	*less*
horizontal	*greater*	*points*

3. Graph each line. Predict whether the slope is negative, positive, or equal to zero. Then check by finding the rise and run of each line.

   (a) $A(-9, 5)$ to $B(-4, 8)$

   (b) $C(-3, 7)$ to $D(0, 2)$

   (c) $E(-10, 2)$ to $F(-4, 1)$

   (d) $G(1, 7)$ to $H(9, 3)$

   (e) $I(0, 0)$ to $J(4, 8)$

   (f) $K(-7, -6)$ to $L(-5, -2)$

   (g) $M(-1, -5)$ to $N(5, -2)$

   (h) $O(5, -5)$ to $P(7, -1)$

4. For each set of points, give the rise from A to B, the run from A to B, and the slope.

  (a) $A(4, 1)$, $B(8, 3)$

  (b) $A(4, 7)$, $B(4, 2)$

  (c) $A(-4, -7)$, $B(1, -7)$

5. Explain why horizontal lines have a slope of 0 and why vertical lines have an undefined slope.

6. Are the points collinear? Justify your answer.

  (a) $A(-3, -4)$, $B(2, 0)$, $C(12, -8)$

  (b) $D(11, 9)$, $E(3, 6)$, $F(7, 5)$

  (c) $G(2, -6)$, $H(-10, -4)$, $I(14, -8)$

7. Graph each line and write the equation.

  (a) Line passes through $(-1, 3)$ and $m = \dfrac{2}{3}$.

  (b) Line passes through $(2, 3)$ and $m = -\dfrac{4}{3}$.

  (c) Line passes through $(3, 2)$ and $m = -\dfrac{3}{2}$.

  (d) Line passes through $(-2, -1)$ and $m = -4$.

8. Find the value of k for each line.
   (a) Line passes through $(2, k)$ and $(3, -2)$, and $m = 2$.

   (b) Line passes through $(-3, 1)$ and $(4, k)$, and $m = \frac{1}{2}$.

   (c) Line passes through $(-8, k)$ and $(2, 3k)$, and $m = -3$.

9. Explain why a slope is expressed as an improper fraction, not as a mixed fraction.

10. If a line passing through $(3, 4)$ has a slope of $-\frac{2}{3}$, state the coordinates of two other points on the line. Choose one point that is lower than $(3, 4)$ and one that is higher.

11. Use two different methods to show that points $A(-1, 4)$, $B(-7, 0)$, and $C(2, 6)$ are collinear.

12. The endpoints on a line segment are $(4, 3)$ and $(6, 4)$. Find a third point on this line segment. (Do not use the midpoint.) Create a graph to check your answer.

13. Create a cross-number puzzle where the answers are slopes or coordinates. (A cross-number puzzle is like a crossword, except the answers are numbers. Use graph paper to set up your puzzle.)

   Check your puzzle by completing it yourself. Then make a copy and exchange puzzles with a classmate.

7.4 Determining the Equation of a Line

The slope-intercept form of an equation is $y = mx + b$, where m represents the slope of the line and b represents the y-intercept. This lesson demonstrated how to generate the **slope-intercept form of the equation** if you are given:

- the graph of the line.
- the slope and the coordinates of one point.
- the coordinates of two points.

Equations can also be written in the standard form $Ax + By + C = 0$. In this form, fractions are usually eliminated, and the coefficient of x is positive.

Example 1

State the equation of line AB.

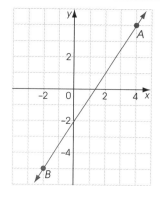

Solution

To write the equation in slope-intercept form, you need to determine the values of m and b in the equation $y = mx + b$.

Step 1

Determine the slope, m, of line AB.

$$m = \frac{rise}{run}$$
$$m = \frac{9}{6} = \frac{3}{2}$$

Step 2

Determine the y-intercept, b, of line AB.

The line crosses the y-axis at the point $(0, -2)$ so the y-intercept is -2. Therefore, $b = -2$.

Step 3

Substitute the values for m and b into the equation $y = mx + b$.

$$y = mx + b$$
$$y = \frac{3}{2}x + (-2)$$
$$y = \frac{3}{2}x - 2$$

Example 2

An electrician charges a flat rate for each house call, plus a certain amount for each hour of service. For 2 h, the bill is $100. For 3 h, the bill is $125.

Find the electrician's flat rate and hourly charge.

Solution

This situation could be shown on a graph where x = number of hours and y = cost.

- The y-intercept (b) is the charge for 0 h of service, or the flat rate.
- The slope (m) indicates the increase in cost (y) each time the number of hours (x) increases by 1. This is the hourly charge.

When you find the equation of the line, you will be able to read the values for m and b.

Step 1

Find two points, (*hours, cost*), on the line.

For 2 h, the cost is $100. Let point A be $(2, 100)$.
For 3 h, the cost is $125. Let point B bc $(3, 125)$.

Step 2

Determine the slope of line AB.

Let $A(2, 100)$ be point 1 and $B(3, 125)$ be point 2.

$$m = \frac{y_2 - y_1}{x_2 - x_1} = \frac{125 - 100}{3 - 2} = 25$$

The hourly rate is $25/h.

Step 3

Use the slope formula to find the equation of the line, then determine b, the flat rate.

Since you want to find the relationship between y and x, substitute (x, y) for (x_2, y_2) and a pair of known coordinates, such as $(2, 100)$, for (x_1, y_1).

$$25 = \frac{y_2 - y_1}{x_2 - x_1}$$
$$25 = \frac{y - 100}{x - 2} \quad \text{Multiply both sides by } x - 2.$$
$$25(x - 2) = y - 100 \quad \text{Simplify the left side.}$$
$$25x - 50 = y - 100 \quad \text{Add 100 to both sides.}$$
$$25x + 50 = y$$
$$y = 25x + 50$$

$b = 50$, so the charge for 0 h is $50. The electrician charges a $50 flat rate, plus $25/h.

Exercises

1. Write sentences that incorporate two or more of these terms. (Use the terms as they relate to linear equations.) Continue until you have used all the terms at least once.

   slope-intercept form   *x-coordinate*   *x-axis*
   linear equation   *y-coordinate*   *y-axis*
   standard form   *x-intercept*   *graph*
   y-intercept

2. Describe each line.
   (a) $x = 4$ **(b)** $y = -1$ **(c)** $x = y$

3. Without drawing the graph, state the slope and the y-intercept for each line.
   (a) $y = 3x + 6$ **(b)** $3x + 2y = 6$

4. Use two different methods to find the y-intercept of the line $3x + y = 8$.

5. Find the equation of each line in standard form.
   (a) The slope is 2 and the line passes through $(4, 3)$.

   (b) The slope is $-\frac{2}{3}$ and the line passes through $(2, 1)$.

   (c) The slope is $\frac{1}{2}$ and the line passes through $(-3, 1)$.

   (d) The slope is -3 and the line passes through $(0, -4)$.

6. Find the standard-form equation for a line passing through each pair of points.
   (a) $(4, 2)$ and $(-1, 3)$ **(b)** $(2, 1)$ and $(-1, -3)$

   (c) $(4, 3)$ and $(8, -1)$ **(d)** $(-2, -3)$ and $(4, 5)$

7. Find the equation for each line.

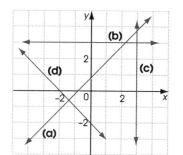

(a)

(b)

(c)

(d)

8. Find the slope-intercept equation of a line through point $(3, -1)$ if:

(a) the slope is $\frac{1}{2}$

(b) the y-intercept is $(0, -2)$

(c) the x-intercept is $(-1, 0)$

(d) the line also passes through $(2, -3)$

9. Find each equation in standard form.

(a) A line passes through $(-4, 2)$ and has the same y-intercept as $2x - y = 3$.

(b) A line has the same x-intercept as $2x + y = 8$ and passes through $(0, 1)$.

(c) A line has the same x-intercept as $3x - y = 9$ and the same y-intercept as $2y - x + 8 = 0$.

Hint: The y-intercept is the value of y when $x = 0$. The x-intercept is the value of x when $y = 0$.

10. $C(5, -9)$ is the center of a circle and $P(8, 2)$ is a point on the circumference. Find the standard-form equation of the line containing radius CP.

11. Determine the area of rectangle $ABCD$.

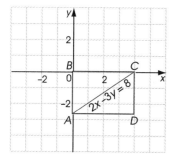

12. A home day care operator charges a weekly flat rate to cover the cost of meals and snacks, plus an hourly fee. One week, the Davidsons paid $42 for 8 h of baby-sitting service, while the Lees paid $170 for 40 h. Determine the weekly flat rate and the hourly fee.

13. Kara works at a bank. The table shows her earnings for one week.

	Mon	Tues	Wed	Thurs	Fri
Hours Worked	3	6	2	5	4
Dollars Earned	19.50	39.00	13.00	32.50	26.00

(a) Graph the data and label the axes.

(b) Find the equation of the line.

(c) What can you conclude about how Kara is paid?

14. Graph six lines on a coordinate grid. Find the equation of each line and then exchange graphs with a classmate to check.

15. Create an equation in slope-intercept form and then create a problem based on your equation. Exchange problems with a classmate.

8 LINEAR SYSTEMS

8.1 Solving Linear Systems by Graphing

In a **system** of linear equations, two or more linear equations are graphed on the same coordinate plane. The **solution** is the point at which the graphs intersect. One way to find this point is to make a table of values for each equation and then look for a point that is common to both tables. A more efficient method is to graph the equations and find the intersection point. A system of linear equations can be **independent** (one point of intersection), **inconsistent** (the lines are parallel, so there are no intersection points), or **dependent** (the lines are collinear, so there are an infinite number of intersection points).

You examined these methods for graphing systems of equations:
- Create a table of values for each equation. Plot and join two points to graph each line.
- Write each equation in the form $y = mx + b$. Plot the y-intercept (b) for one equation, and then use the slope (m) to identify another point on the line. Repeat with the other equation.
- Identify the x- and y-intercepts of each line. To find the x-intercept, let $y = 0$ in the equation. To find the y-intercept, let $x = 0$. Plot each line by joining the x- and y-intercepts.

Example

A videocassette store offers two different monthly payment plans for employees. With Plan A, the employee receives a $1,500 salary plus 25¢ for each tape sold. With Plan B, the employee receives no salary but earns 50¢ for each tape sold.

How many tapes would an employee need to sell in a month before Plan B would start to pay more than Plan A?

Solution

The monthly pay (y) depends on the number of tapes sold (x).

Plan A can be expressed as $y = 1,500 + 0.25x$.

Plan B can be expressed as $y = 0.5x$.

To find the point where Plan B begins to pay more, locate the intersection point of the graphs. One way to graph each equation is to use the form $y = mx + b$, where b represents the y-intercept and m represents the slope.

$y = mx + b$

$y = 0.25x + 1,500$ (Plan A) The y-intercept is at (0, 1,500). The slope is 0.25.

$y = 0.5x + 0$ (Plan B) The y-intercept is at (0, 0). The slope is 0.5.

Plan A

To graph $y = 0.5x + 0$, plot the y-intercept at (0, 0). The slope is 0.5 or $\frac{5}{10}$ or $\frac{500}{1,000}$, so the graph rises 500 units for every 1,000 units of run. Plot a second point so it is 1,000 units to the right and 500 units up, at (1,000, 500). Join the points to graph the line.

Plan B

To graph $y = 0.25x + 1,500$, plot the y-intercept at (0, 1,500). The slope is 0.25 or $\frac{25}{100}$ or $\frac{250}{1,000}$, so the graph rises 250 units for every 1,000 units of run. Plot a second point so it is 1,000 units to the right of the y-intercept and 250 units up, at (1,000, 1,750). Join the points to graph the line.

The lines intersect at (6,000, 3,000). This is the solution to the system. It means that an employee would earn $3,000 under either payment plan if 6,000 tapes were sold. Plan B will pay more than Plan A if an employee sells more than 6,000 tapes in a month.

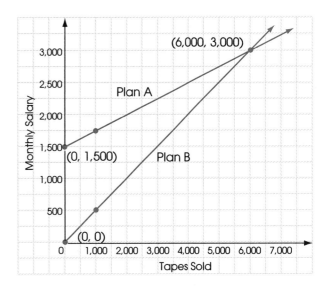

Exercises

1. Explain.

   (a) What is a system of linear equations?

   (b) How can you solve a system of linear equations by graphing?

   (c) How can you verify your solution?

2. Sketch the three different types of linear systems that can be produced by two equations. Identify each system type by name.

3. Create a table of values for each system to determine its solution. Do not graph the system.

   (a) $y = -3x - 4$ and $y = -10$

   (b) $y = \dfrac{2x}{5} + 6$ and $x + 5 = 0$

   (c) $y = (x - 3) \div 4$ and $-2y = 4x + 6$

4. Solve each system by graphing. Create a table of values to help you graph each line.
   (a) $x + y - 9 = 0$ and $y = 2x$

   (b) $5x - 4y + 24 = 0$ and $y = -\frac{x}{4}$

   (c) $3x + 4y = 12$ and $-6x - 8 = 8y$

5. Use the slope and y-intercept to graph each equation. Then use your graph to solve each system.

   (a) $3y = 3x - 6$ and $9 = 8x - 9y$

   (b) $x - 4 = 0$ and $y = 7$

   (c) $5x + 8y = 48$ and $6 - y = \frac{5x}{8}$

6. Use the x- and y-intercepts to graph each equation. Then use your graph to solve the system.
   (a) $7 - \frac{7x}{9} = y$ and $-9y = 7x - 63$ (b) $y = -\frac{x}{2} + 2$ and $x - y = 7$

   (c) $2y + 3x = 8$ and $x + y = 5$ (d) $y = (2x - 3) \div 4$ and $1 = x - y$

7. Use a method of your choice to graph and solve each system.
   (a) $y = 3x - 5$ and $x + 3y - 15 = 0$ (b) $5y + 2x = 0$ and $4x + 3y = 14$

8. Solve each system. How accurate are the solutions? Explain.
   (a) $y = \frac{x}{4}$ and $3x + 2y + 7 = 0$ (b) $9x + 4y + 13 = 0$ and $3x - 2y - 9 = 0$

9. Is (–2, –5) the solution to each system? Verify.

    (a) $y = 4x + 3$ and $y = (x - 8) \div 2$ **(b)** $3x - 4y - 14 = 0$ and $5x - 2y = 0$

10. Before you graph a system of linear equations, how can you predict whether it will have no solution, one solution, or an infinite number of solutions?

11. Another line, $y = mx + b$, will be added to this graph to create a system. Suggest values for m and b that would give the system:

    (a) one solution

    (b) no solution

    (c) an infinite number of solutions

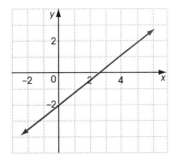

12. The system includes $y = \dfrac{3x}{4} - 2$ (graphed in Problem 11) and $-2x + dy - 6 = 0$. Are there values for d that will form each type of system? Give examples.

    (a) a dependent system

    (b) an independent system

    (c) an inconsistent system

13. Two ants are crawling in straight lines across a coordinate grid. The first ant travels along a path described by $y = \dfrac{5x}{2} - 13$. The second ant travels along a path described by $y = -6x + 4$. Where will their paths cross?

14. Airplane A travels along a path described by $y = \dfrac{4x}{5} + 12$. Airplane B travels at a higher altitude, along a path described by $5y = 4x + 60$. Will Airplane A fly directly below Airplane B? Explain.

15. A diamond is placed with its center at (0, 0). A laser will slice cleanly through the diamond if it is placed in quadrant I at an angle of 45° from the x-axis. Suggest some ordered pairs that lie along the path of the beam.

8.2 Solving Linear Systems by Elimination

In this lesson, you have
- solved systems of equations algebraically using **substitution** and **elimination**.
- verified **solutions** to systems.
- determined, algebraically, systems that are **independent**, **dependent**, or **inconsistent**.
- solved problems using **systems of equations**.

Example 1

Solve the following system using substitution.

❶ $\qquad 3x - 4y = 11$
❷ $\qquad\qquad x = 2y + 3$

Solution

Substitute the expression $2y + 3$ for x from equation ❷ into equation ❶.

❶ $3(2y + 3) - 4y = 11$
$\qquad 6y + 9 - 4y = 11$
$\qquad\qquad\quad 2y = 2$
$\qquad\qquad\quad\ y = 1$

Solve for x using equation ❷.

$$x = 2(1) + 3$$
$$x = 5$$

The solution is $(5, 1)$.

Example 2

Solve using elimination.

❶ $\qquad \dfrac{x-1}{3} + \dfrac{y-2}{4} = 1$
❷ $\qquad 2x - 5y + 15 = 0$

Solution

Rearrange equation ❶ in the form $Ax + By = C$.

❶ $12\left(\dfrac{x-1}{3}\right) + 12\left(\dfrac{y-2}{4}\right) = 12(1)$ $\quad$ Multiply by 12.
$\qquad 4(x-1) + 3(y-2) = 12$
$\qquad\quad 4x - 4 + 3y - 6 = 12$
$\qquad\qquad\qquad 4x + 3y = 22$

Rearrange equation ❷ in this form to give this system:

❶ $\qquad 4x + 3y = 22$
❷ $\qquad 2x - 5y = -15$

To eliminate, multiply equation ❷ by 2 and subtract.

❶ $\qquad 4x + 3y = 22$
❷ $\qquad \underline{4x - 10y = -30}$
$\qquad\qquad 13y = 52 \qquad\qquad$ Subtract.
$\qquad\qquad\ \ y = 4$

To find x, substitute 4 for y in equation ❷.
$$2x - 5(4) = -15$$
$$2x - 20 = -15$$
$$2x = 5$$
$$x = \frac{5}{2} \text{ or } 2.5$$

The solution is $(2.5, 4)$.

Example 3

At Busy Baker Foods, 4 lb of Swiss cheese and 5 lb of Gouda cheese cost $12.00. Six pounds of Swiss and nine pounds of Gouda cost $20.10. Find the price per pound for each type of cheese.

Solution

Let s be the price per pound for the Swiss cheese and g be the price per pound for the Gouda cheese, in dollars.

Therefore,

❶ $\quad 4s + 5g = 12$
❷ $\quad 6s + 9g = 20.10$

Solve by elimination. Multiply equation ❶ by 3 and equation ❷ by –2 and add.

❶ $\quad\quad 12s + 15g = 36$
❷ $\quad -12s - 18g = -40.20$
$$-3g = -4.2$$
$$g = 1.4$$

Substitute $g = 1.4$ into equation ❶ and solve for s.
$$4s + 5(1.4) = 12$$
$$4s + 7 = 12$$
$$4s = 5$$
$$s = 1.25$$

Therefore, 1 lb of Swiss cheese costs $1.25, and 1 lb of Gouda costs $1.40.

Exercises

1. Solve the following systems by substitution.

   (a) $x - y = 1$ and $y = -x + 5$

   (b) $3x - 4y = 8$ and $y = 17 - 4x$

   (c) $y - 3x + 6 = 0$ and $9x - 4y = 3$

2. Solve the following systems by elimination.

    (a) $4a + 12b = 7$ and $4a - 8b = 9$

    (b) $3x - 5y = 1$ and $6(x - 3) = 10(y + 1)$

    (c) $\dfrac{x + y}{3} + \dfrac{x - y}{5} = -2$ and $3x - 2y = 13$

3. Solve by any method. (Hint: Let $x = \dfrac{1}{a}$ and $y = \dfrac{1}{b}$ first.)

    $\dfrac{1}{a} + \dfrac{4}{b} = 3$ and $\dfrac{3}{a} - \dfrac{2}{b} = -5$

4. This system is inconsistent. Find A.

    $2x + 5y = 8$
    $Ax + 4y = 0$

5. The points $(-3, 4)$ and $(2, 1)$ lie on the line $Ax + By = 11$. Find A and B.

6. This system is dependent. Find A and B.

$$3x + 4y = 5$$
$$Ax + 4y = B$$

7. The lines $3x - 8y + 39 = 0$, $4x + y - 18 = 0$, and $2x + 4y - 2 = 0$ intersect and form a triangle. Find the coordinates of the vertices of the triangle.

8. Anita invested a total of \$5,000 at the beginning of the year in two different funds. At the end of the year, her investment had grown to \$5,531. The money in the first fund earned 9%, while the money in the second fund earned 13.5%. How much of the \$5,000 was invested in each fund at the beginning of the year? Use a system of equations to solve this problem.

9. Oakdale High School is holding a dance on Friday. The student council has sold tickets to younger students at \$4.50 each and to senior students at \$6 each. Unfortunately, the council member in charge of selling tickets forgot to keep a record of the number of tickets sold at each price. However, he does know that a total of \$792 was collected, and 152 tickets were sold all together. Set up and solve a system of equations to determine the number of tickets sold at each price.

8.3 Solving Three Equations in Three Variables

In this lesson, you learned how to use algebraic techniques to solve a system of three linear equations in three variables. The solution, if one exists, is written as an ordered triple (x, y, z). It is possible for a system to have no solutions, or to have an infinite number of solutions.

Example

Solve:

❶ $3x + y + 2z = 14$
❷ $2x + y - z = 5$
❸ $x - 4y + 2z = 3$

Solution

Step 1

Choose a variable to eliminate. In this system, the equations $3x + y + 2z = 14$ and $x - 4y + 2z = 3$ both contain the term $2z$. If you multiply $2x + y - z = 5$ by 2, the resulting equation, $4x + 2y - 2z = 10$, will also contain the term $2z$. Eliminate the z-terms.

Step 2

To eliminate the z-terms from one pair of equations, add ❶ to $2 \times$ ❷.

❶ $3x + y + 2z = 14$
❷ ×2 $\underline{4x + 2y - 2z = 10}$
❹ $7x + 3y = 24$

Multiply equation ❷ by 2 to change $-z$ to $-2z$. Add the equations to eliminate the z-terms.

Step 3

Eliminate the z-term from a different pair of equations.

❷ ×2 $4x + 2y - 2z = 10$
❸ $\underline{x - 4y + 2z = 3}$
❺ $5x - 2y = 13$

Step 4

You have created two 2-variable equations, ❹ and ❺, where x and y have the same values as in the original system. To solve this new system, eliminate one of the variables from equations ❹ and ❺.

❹ ×2 $14x + 6y = 48$
❺ ×3 $\underline{15x - 6y = 39}$
                   $29x = 87$
                   $x = 3$

There are no common terms in equations ❹ and ❺. However, the y-terms have an obvious common multiple, $6y$. Multiply ❹ by 2 and ❺ by 3 to get $6y$ in both equations. Add the two equations to eliminate the y-terms and solve for x.

Step 5

Substitute $x = 3$ into equation ❹ or ❺ to find the value for y.

❺ $5x - 2y = 13$
                   $5(3) - 2y = 13$
                   $-2y = -2$
                   $y = 1$

Step 6

Substitute $x = 3$ and $y = 1$ into equation ❶, ❷, or ❸ to find the value for z.

❷ $2x + y - z = 5$
                   $2(3) + 1 - z = 5$
                   $-z = -2$
                   $z = 2$

Step 7

Write a final statement.

The solution for this system is $x = 3$, $y = 1$, and $z = 2$ or $(3, 1, 2)$.

Exercises

1. Show the steps you would use to reduce this system of equations to a single equation with two variables.

   $5a + 4b - c = 10$

   $3a - 6b + 2c = 15$

2. Why is it useful to know how to eliminate a variable from a set of equations?

3. Determine if the ordered triple is a solution for the system.

   (a) $(2, -1, 3)$ for

   $3x + 2y + 2z = 7$

   $x - y - z = 0$

   $2x - 3y + z = 10$

   (b) $(4, 2, 0)$ for

   $2x - y + 3z = 6$

   $x + y - 4z = 6$

   $3x - 2z = 6$

   (c) $(-10, -9, 5)$ for

   $2x - 5y - 4z = 5$

   $5x - 2y + 2z = -22$

   $x + 4y - 3z = -2$

   (d) $(-2, 5, 3)$ for

   $x - y = -7$

   $x - z = -5$

   $x + y + z = 6$

4. Find the errors in the first part of this solution. Then continue the solution to find values for a, b, and c.

   Problem

   Solve: ❶ $2a + b - 4c = 17$

   ❷ $5a - 2b = 27$

   ❸ $b + 3c + 7 = 0$

   Solution

   ❶ $\qquad 2a + b - 4c = 17$

   ❷ $- \underline{\quad b + 3c + 7 = 0}$

   ❸ $\qquad 2a\ -c + 7 = 17$

5. Solve each system of equations.

(a) $x + 2y + z = 3$
$2x - y - z = 4$
$-x - y + 2z = -5$

(b) $3x + 7y - 2z = 10$
$x - y + 6z = 11$
$2x + 3y + 2z = -15$

(c) $2x - 3y + z = 0$
$x + 4y + 5z = 7$
$4x + 5y + 6z = -6$

(d) $x + 2y - z = 3$
$2x - 3y + 3z = 0$
$y - 2x = -8$

(e) $2x + y + 3z = 6$
$5y + 3z = 7$
$4x - 3y + 3z = 5$

(f) $3y + z = 9$
$2x + y + 3z = -1$
$4x - 5y = -12$

6. Sam has $440 in $5, $10, and $20 bills. He has 38 bills in all. If you add the number of $10 bills to the number of $20 bills, and then subtract 10, the result is the number of $5 bills. How many bills of each denomination does he have?

7. Celine works twice as many hours as she spends in class. She also studies 3 hours for every hour in class. If she spends 60 hours per week on work and school, how many hours does she spend on each activity: attending class, studying, and working?

8. An ounce of macadamia nuts (salted and roasted in oil) contains 222 calories. There are twice as many grams of carbohydrates as protein. The total weight of the protein, fat, and carbohydrates is 28 g. There are 4 calories per gram of protein and carbohydrates and 9 calories per gram of fat. How many grams are there of each: carbohydrates, fat, and protein?

9. A group invests \$50,000 in three stocks. The first stock pays 5%, the second pays 7%, and the third pays 4%. The total interest earned from the three stocks is \$2,810. If \$6,000 less than three times the second investment is four times the first investment, how much was invested at each rate of interest?

10. Write any ordered triple (a, b, c) where all three values are integers. Use these values to develop three equations where a, b, and c represent the values you chose. For example, if the numbers are (1), (2), and (3), then one equation might be $4(1) + 2(2) - 5(3) = -7$, or $4a + 2b - 5c = -7$.

Solve your system of three equations to make sure the result is the triple you started with. Then trade systems with a classmate and solve again.

8.4 Solving Linear Systems by Matrices

In this lesson, you learned how to solve linear systems using matrices. The three elementary row operations are:

Type 1: Any two rows of a matrix can be interchanged.

Type 2: Any row of a matrix can be multiplied by a non-zero constant.

Type 3: Any row of a matrix can be changed by adding a non-zero constant multiple of another row to it.

Example 1

Solve the system of two linear equations. Verify your solution.

$$\begin{cases} 3x + 6y = 6 \\ 2x - 2y = 0 \end{cases}$$

Solution

Write the system as an augmented matrix and then use elementary row operations to simplify the matrix.

$$\begin{bmatrix} 3 & 6 & | & 6 \\ 2 & -2 & | & 0 \end{bmatrix}$$

Write the system as an augmented matrix and then use elementary row operations to simplify the matrix.

$$= \begin{bmatrix} 1 & 2 & | & 2 \\ 1 & -1 & | & 0 \end{bmatrix} \quad \begin{array}{l} \frac{1}{3} \times R_1 \\ \frac{1}{2} \times R_2 \end{array}$$

$$= \begin{bmatrix} 1 & 2 & | & 2 \\ 0 & -3 & | & -2 \end{bmatrix} \quad R_2 - R_1$$

$$= \begin{bmatrix} 1 & 2 & | & 2 \\ 0 & 1 & | & \frac{2}{3} \end{bmatrix} \quad \left(\frac{-1}{3}\right) \times R_2$$

Write the resulting matrix as a system of two equations.

$$\begin{cases} x + 2y = 2 \\ y = \frac{2}{3} \end{cases}$$

Substitute $y = \frac{2}{3}$ into equation 1 to solve for x.

$$x + 2\left(\frac{2}{3}\right) = 2$$

$$x + \frac{4}{3} = 2$$

$$x = \frac{2}{3}$$

The solution is $\left(\frac{2}{3}, \frac{2}{3}\right)$.

Verify the solution in the original system.

Equation 1

L.S.	R.S.
$3x + 6y$	6
$= 3\left(\frac{2}{3}\right) + 6\left(\frac{2}{3}\right)$	
$= 2 + 4$	
$= 6$	

Equation 2

L.S.	R.S.
$2x - 2y$	0
$= 2\left(\frac{2}{3}\right) - 2\left(\frac{2}{3}\right)$	
$= \left(\frac{4}{3}\right) - \left(\frac{4}{3}\right)$	
$= 0$	

Example 2

Solve the system of three linear equations.

$$\begin{cases} x - 2y - 2z = -1 \\ 4x - y + 3z = 4 \\ x + 5y + 9z = 7 \end{cases}$$

Solution

$$\begin{bmatrix} 1 & -2 & -2 & | & -1 \\ 4 & -1 & 3 & | & 4 \\ 1 & 5 & 9 & | & 7 \end{bmatrix}$$

Write the system as an augmented matrix and then use elementary row operations to simplify the matrix.

$$= \begin{bmatrix} 1 & -2 & -2 & | & -1 \\ 0 & 7 & 11 & | & 8 \\ 0 & 7 & 11 & | & 8 \end{bmatrix} \quad \begin{array}{l} -4R_1 + R_2 \\ -R_1 + R_3 \end{array}$$

$$= \begin{bmatrix} 1 & -2 & -2 & | & -1 \\ 0 & 7 & 11 & | & 8 \\ 0 & 0 & 0 & | & 0 \end{bmatrix} \quad -R_2 + R_3$$

Write the resulting matrix as a system of three equations.

$$\begin{cases} x - 2y - 2z = -1 \\ 0x + 7y + 11z = 8 \\ 0x + 0y + 0z = 0 \end{cases}$$

Equation 3 is $0x + 0y + 0z = 0$. There are infinitely many ordered triples that satisfy equation 3, so this is a *dependent system*.

Exercises

1. Complete the sentences.

   (a) A _____ is a rectangular array of numbers.

   (b) The numbers in a matrix are called its _____.

   (c) A 3×4 matrix has ____ rows and ____ columns.

   (d) Elementary _____ operations are used to produce new matrices that lead to the solution of a system.

   (e) A matrix that represents the equations of a system is called an _____ matrix.

   (f) The augmented matrix $\begin{bmatrix} 1 & 3 & | & -2 \\ 0 & 1 & | & 4 \end{bmatrix}$ has 1s in its _____.

2. Explain how solving a system of equations with an augmented matrix is similar to solving by elimination.

3. For each matrix, state the number of rows and the number of columns.

   (a) $\begin{bmatrix} 4 & 6 & | & -1 \\ \frac{1}{2} & 9 & | & -3 \end{bmatrix}$

   (b) $\begin{bmatrix} 1 & -2 & 3 & | & 1 \\ 0 & 1 & 6 & | & 4 \\ 0 & 0 & 1 & | & \frac{1}{3} \end{bmatrix}$

4. For each augmented matrix, give a system of equations it represents.

   (a) $\begin{bmatrix} 1 & 6 & | & 7 \\ 0 & 1 & | & 4 \end{bmatrix}$

   (b) $\begin{bmatrix} 2 & -2 & 9 & | & 1 \\ 3 & 1 & 1 & | & 0 \\ 2 & -6 & 8 & | & -7 \end{bmatrix}$

5. Write a system of equations represented by the augmented matrix. Then use substitution to find the solution.

   (a) $\begin{bmatrix} 1 & -1 & | & -10 \\ 0 & 1 & | & 6 \end{bmatrix}$

   (b) $\begin{bmatrix} 1 & -2 & 1 & | & -16 \\ 0 & 1 & 2 & | & 8 \\ 0 & 0 & 1 & | & 4 \end{bmatrix}$

6. Matrices were used to solve a system of linear equations. The final matrix is shown here. Explain what the result tells about the system.

   (a) $\begin{bmatrix} 1 & 3 & | & 4 \\ 0 & 0 & | & 7 \end{bmatrix}$

   (b) $\begin{bmatrix} 1 & 2 & | & -4 \\ 0 & 0 & | & 0 \end{bmatrix}$

7. Consider the matrix $\begin{bmatrix} -3 & 1 & | & -6 \\ 1 & -4 & | & 4 \end{bmatrix}$

   (a) Explain what is meant by $R_1 \leftrightarrow R_2$. Then perform the operation on the matrix.

   (b) Explain what is meant by $3R_1 + R_2$. Then perform the operation on the answer to part (a).

8. Use matrices to solve each system of equations.

   (a) $\begin{cases} x + y = 2 \\ x - y = 0 \end{cases}$
   (b) $\begin{cases} x + y = 3 \\ x - y = -1 \end{cases}$
   (c) $\begin{cases} 2x + y = 1 \\ x + 2y = -4 \end{cases}$

   (d) $\begin{cases} 5x - 4y = 10 \\ x - 7y = 2 \end{cases}$
   (e) $\begin{cases} 2x - y = -1 \\ x - 2y = 1 \end{cases}$
   (f) $\begin{cases} 2x - y = 0 \\ x + y = 3 \end{cases}$

   (g) $\begin{cases} 3x + 4y = -12 \\ 9x - 2y = 6 \end{cases}$
   (h) $\begin{cases} 2x - 3y = 16 \\ -4x + y = -22 \end{cases}$
   (i) $\begin{cases} x + y + z = 6 \\ x + 2y + z = 8 \\ x + y + 2z = 9 \end{cases}$

   (j) $\begin{cases} x - y + z = 2 \\ x + 2y - z = 6 \\ 2x - y - z = 3 \end{cases}$
   (k) $\begin{cases} 3x + y - 3z = 5 \\ x - 2y + 4z = 10 \\ x + y + z = 13 \end{cases}$
   (l) $\begin{cases} 2x + y - 3z = -1 \\ 3x - 2y - z = -5 \\ x - 3y - 2z = -12 \end{cases}$

   (m) $\begin{cases} 3x - 2y + 4z = 4 \\ x + y + z = 3 \\ 6x - 2y - 3z = 10 \end{cases}$
   (n) $\begin{cases} 3a + 2b + c = 8 \\ 6a - b + 2c = 16 \\ -9a + b - c = -20 \end{cases}$

9. Use matrices to solve each system of equations. Indicate whether the system is independent, dependent, or inconsistent.

   (a) $\begin{cases} x - 3y = 9 \\ -2x + 6y = 18 \end{cases}$
   (b) $\begin{cases} -6x + 12y = 10 \\ 2x - 4y = 8 \end{cases}$
   (c) $\begin{cases} 4x + 4y = 12 \\ -x - y = -3 \end{cases}$

(d) $\begin{cases} 5x - 15y = 10 \\ 2x - 6y = 4 \end{cases}$

(e) $\begin{cases} 6x + y - z = -2 \\ x + 2y + z = 5 \\ 5y - z = 2 \end{cases}$

(f) $\begin{cases} 2x + 3y - 2z = 18 \\ 5x - 6y + z = 21 \\ 4y - 2z = 6 \end{cases}$

(g) $\begin{cases} 2x + y - z = 1 \\ x + 2y + 2z = 2 \\ 4x + 5y + 3z = 3 \end{cases}$

(h) $\begin{cases} x - 3y + 4z = 2 \\ 2x + y + 2z = 3 \\ 4x - 5y + 10z = 7 \end{cases}$

(i) $\begin{cases} 5x + 3y = 4 \\ 3y - 4z = 4 \\ x + z = 1 \end{cases}$

(j) $\begin{cases} y + 2z = -2 \\ x + y = 1 \\ 2x - z = 0 \end{cases}$

(k) $\begin{cases} x - y = 1 \\ 2x - z = 0 \\ 2y - z = -2 \end{cases}$

(l) $\begin{cases} x + y - 3z = 4 \\ 2x + 2y - 6z = 5 \\ -3x + y - z = 2 \end{cases}$

10. Georgette has a total of $155 consisting of two-dollar and five-dollar bills. If she has 40 bills in all, how many of each does she have?

11. The cash receipts from two sold-out performances of a play are shown with the ticket prices. Find the number of seats in each of the three sections of the 800-seat theater.

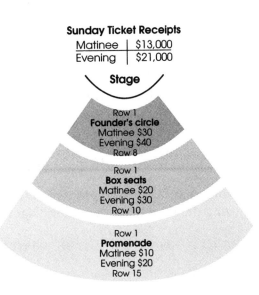

Sunday Ticket Receipts

| Matinee | $13,000 |
| Evening | $21,000 |

Stage

Row 1
Founder's circle
Matinee $30
Evening $40
Row 8

Row 1
Box seats
Matinee $20
Evening $30
Row 10

Row 1
Promenade
Matinee $10
Evening $20
Row 15

In this lesson, you learned how to calculate determinants and use them to solve systems of equations.

- **Value of a 2 × 2 determinant:** For the matrix $\begin{vmatrix} a & b \\ c & d \end{vmatrix}$, the value of the determinant is $\begin{vmatrix} a & b \\ c & d \end{vmatrix} = ad - bc.$

- **Value of a 3 × 3 determinant:** $\begin{vmatrix} a_1 & b_1 & c_1 \\ a_2 & b_2 & c_2 \\ a_3 & b_3 & c_3 \end{vmatrix} = a_1 \begin{vmatrix} b_2 & c_2 \\ b_3 & c_3 \end{vmatrix} - b_1 \begin{vmatrix} a_2 & c_2 \\ a_3 & c_3 \end{vmatrix} + c_1 \begin{vmatrix} a_2 & b_2 \\ a_3 & b_3 \end{vmatrix}$

- **Cramer's Rule for two equations in two variables:** The solution of the system $\begin{cases} ax + by = e \\ cx + dy = f \end{cases}$ is given by

$$x = \frac{D_x}{D} = \frac{\begin{vmatrix} e & b \\ f & d \end{vmatrix}}{\begin{vmatrix} a & b \\ c & d \end{vmatrix}} \text{ and } y = \frac{D_y}{D} = \frac{\begin{vmatrix} a & e \\ c & f \end{vmatrix}}{\begin{vmatrix} a & b \\ c & d \end{vmatrix}}$$

- **Cramer's Rule for three equations in three variables**

The solution of the system $\begin{cases} ax + by + cz = j \\ dx + ey + fz = k \\ gx + hy + iz = l \end{cases}$ is given by $x = \frac{D_x}{D}$ and $y = \frac{D_y}{D}$ and $z = \frac{D_z}{D}$

Example

Solve, using Cramer's Rule.

$$x + 2y = 4$$
$$5x - y = -13$$

Solution

Form the augmented matrix.

$$\begin{bmatrix} 1 & 2 & | & 4 \\ 5 & -1 & | & -13 \end{bmatrix}$$

By Cramer's Rule, $x = \dfrac{D_x}{D}$ and $y = \dfrac{D_y}{D}$.

To find D_x, substitute the constant values for the x-values.

To find D_y, substitute the constant values for the y-values.

$$D = \begin{vmatrix} 1 & 2 \\ 5 & -1 \end{vmatrix} \qquad D_x = \begin{vmatrix} 4 & 2 \\ -13 & -1 \end{vmatrix} \qquad D_y = \begin{vmatrix} 1 & 4 \\ 5 & -13 \end{vmatrix}$$

Evaluate each determinant by calculating the product of the main diagonal entries and then subtracting the product of the other diagonal entries.

$D = (1)(-1) - (5)(2) \qquad D_x = (4)(-1) - (-13)(2) \qquad D_y = (1)(-13) - (5)(4)$

$D = -1 - 10 \qquad D_x = -4 + 26 \qquad D_y = -13 - 20$

$D = -11 \qquad D_x = 22 \qquad D_y = -33$

Use Cramer's Rule to find values for x and y.

$$x = \frac{D_x}{D} \qquad y = \frac{D_y}{D}$$
$$= \frac{22}{-11} \qquad = \frac{-33}{-11}$$
$$= -2 \qquad = 3$$

Therefore, $(x, y) = (-2, 3)$.

Exercises

1. Define each term.
   (a) row   (b) column

   (c) augmented matrix   (d) square matrix

   (e) minor

2. Tell how to find the minor of an element of a determinant.

3. Tell how to find x when solving a system of three linear equations by Cramer's Rule.

4. Evaluate the determinants.
   (a) $\begin{vmatrix} 2 & 3 \\ -2 & 1 \end{vmatrix}$ (b) $\begin{vmatrix} 3 & -2 \\ -2 & 4 \end{vmatrix}$

   (c) $\begin{vmatrix} -1 & 4 \\ -3 & -2 \end{vmatrix}$ (d) $\begin{vmatrix} -1 & -2 \\ -2 & -4 \end{vmatrix}$

5. Complete the solution.

Problem

Evaluate.

$$\begin{vmatrix} 1 & 2 & -1 \\ 3 & 0 & -2 \\ 3 & -1 & 2 \end{vmatrix}$$

Solution

Expand the minors about the second row. Use the second row because it has a 0, which will help you eliminate some of the terms. The second column could also be used to get the same result.

Use the second row of the array of signs $\begin{vmatrix} + & - & + \\ - & + & - \\ + & - & + \end{vmatrix}$.

$D = -3\begin{vmatrix} 2 & -1 \\ -1 & 2 \end{vmatrix} + 0\begin{vmatrix} 1 & -1 \\ 3 & 2 \end{vmatrix} - (-2)\begin{vmatrix} 1 & 2 \\ 3 & -1 \end{vmatrix}$ Find the minor of each element in the second row.

6. Evaluate the determinants.

(a) $\begin{vmatrix} 1 & 2 & 1 \\ 1 & 2 & -1 \\ 2 & 1 & 2 \end{vmatrix}$

(b) $\begin{vmatrix} 3 & 5 & 1 \\ 6 & -2 & 2 \\ 8 & -1 & 4 \end{vmatrix}$

(c) $\begin{vmatrix} 1 & -4 & 7 \\ 2 & 5 & 8 \\ 3 & 6 & 9 \end{vmatrix}$

7. Solve each system of equations using Cramer's Rule. Identify dependent or inconsistent systems.

(a) $x - y = -4$
    $2x + 3y = 7$

(b) $y = 2x + 5$
    $3x - 5y = -4$

(c) $0.1x + 0.2y = 1.1$
    $\dfrac{2x - y}{2} = 1$

(d) $x + y + z = 4$
    $2x + y - z = 1$
    $2x - 3y + z = 1$

(e) $2x + 2y + 3z = 10$
$3x + y - z = 0$
$x + y + 2z = 6$

(f) $2x + y - z = 1$
$x + 2y + 2z = 2$
$4x + 5y + 3z = 3$

(g) $2x + 3y + 4z = 6$
$2x - 3y - 4z = -4$
$4x + 6y + 8z = 12$

(h) $x + y = 1$
$\frac{1}{2}y + z = \frac{5}{2}$
$x - z = -3$

8. A system of sending signals uses two flags held in various positions to represent letters of the alphabet. The illustration below shows how the letter U is signaled. Find x and y if y is 30° more than x.

9. An investor wants to average a 6.6% return by investing \$20,000 in the three stocks listed in the chart. Because HiTech is a high-risk investment, he wants to invest three times as much in SaveTel and OilCo combined as he invests in HiTech. How much should he invest in each?

Stock	Rate of Return
HiTech	10%
SaveTel	5%
OilCo	6%

10. For a system of equations, $D = -12$, $x = 3$, and $y = -2$. Find D_x and D_y.

11. Explain why it is easier to evaluate a determinant when the row or column that is expanded by minors contains a zero.

8.6 Solving Systems of Linear Inequalities

You have solved problems involving linear systems using the following five-step method:

Think about the problem.
Make a plan.
Solve the problem.
Look back.
Look ahead.

Example

A jeweler sells gold ring bands and diamond rings. Her store display cabinet cannot hold more than 100 rings. The average cost of a gold band is $200, and the average cost of a diamond ring is $2,000. The appraised value of the entire stock of rings is at least $100,000. Represent the situation graphically and shade the region containing the possible combinations of the number of each type of ring.

Solution

Think about the problem.
Identify the given information.

The total number of rings is less than or equal to 100.

The average cost of a gold band is $200.

The average cost of a diamond ring is $2,000.

The appraised value of the entire stock of rings is greater than or equal to $100,000.

Make a plan.
State the variables and label the axes of the graph.

Create the linear inequalities using the information in the problem.

Graph the system of inequalities.

Determine the solution of the system of inequalities.

Solve the problem.
1. State the variables and label the axes of the graph.
   Let g = number of gold ring bands
   Let d = number of diamond rings

2. Create the linear inequalities using the information in the problem.

$$g + d \leq 100$$
$$200g + 2{,}000d \geq 100{,}000$$

3. Graph the system of inequalities.

$$\begin{cases} d \leq -g + 100 \\ d \geq -\dfrac{1}{10}g + 50 \end{cases}$$

4. Determine the solution of the system of inequalities.

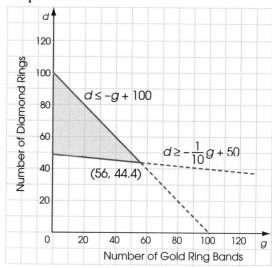

Look back and look ahead.
Determine the maximum number of gold rings that the jeweler could have in stock. The solution region illustrates that the maximum possible gold ring bands within the display cabinet could be 56.

$$200(56) + 2{,}000(44) = 99{,}200$$
$$< 100{,}000$$

Try 55.
$$200(55) + 2{,}000(45) = 101{,}000$$

The maximum number of gold rings is 55.

Exercises

1. Claudine is planning to make meat pies. At her local supermarket, ground pork costs $4/lb, and ground beef costs $5/lb. Claudine does not plan to spend more than $20 on meat for the meat pies. The shaded region represents all of the possible combinations of pork and beef that Claudine can use to make meat pies.

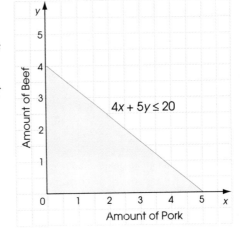

   Determine if the statement is true or false.

   (a) If Claudine purchases 4 lb of ground beef, she can buy no ground pork.

   (b) If Claudine purchases 4 lb of ground pork, she can buy no ground beef.

   (c) If Claudine purchases 2 lb of ground pork, she can buy 3 lb of ground beef.

   (d) If Claudine wants to purchase equal amounts of ground beef and ground pork, the most she could buy is 2.5 lb of each.

2. A record store sold more than 250 discounted compact discs and tapes in one day. The compact discs sold for $10, and the tapes sold for $5. The total cash receipts on the discounted items did not exceed $1,750.

   Define the variables and write two inequalities to represent the given information.

   Graph the solution of the system of inequalities to show the possible combinations of compact discs and tapes that could have been sold that day.

3. Schneider's Supermarket sells sugar in 2 lb and 4 lb bags. One day it sold more than 200 lb but less than 300 lb of sugar. Determine if 80 four-pound bags and 20 two-pound bags is a possible combination of each type of bag sold.

4. A local grocery store sells newspapers and magazines. The average cost of a newspaper is $1, and the average cost of a magazine is $3. Each week the store sells at least twice as many newspapers as magazines and over $500 worth of periodicals altogether.

Define the variables and write two inequalities to represent the given information.

Graph the solution of the system of inequalities to show the possible combinations of newspapers and magazines that could have been sold that week.

5. Albert and Kim are planning to build a rectangular deck. To accommodate their lawn furniture, they estimate that the length of the deck must be greater than the width, the perimeter must be at least 20 yd, and the area must be at least 30 yd^2.

Define the variables and write three inequalities to represent the given information.

Graph the solution of the system of inequalities to show the possible combinations of dimensions of the deck.

Determine if a deck that has a width of 4 yd and a length of 5 yd is a possible deck design for Albert and Kim to build. Justify your answer.

6. A vending machine accepts only dimes and quarters. At most, the machine can hold 100 coins. When the machine is emptied at the end of the day, there is at least $20 worth of coins.

Define the variables and write two inequalities to represent the given information.

Graph the solution of the system of inequalities to show the possible combinations of dimes and quarters that could have been in the machine.

7. Create a problem containing systems of inequalities. Illustrate the steps required to solve the problem. Analyze each step and describe where a mistake is likely to be made. Rewrite that step and include the error. Exchange your question with your classmates. Ask them to find the step that contains the first error and to solve the problem correctly.

In this lesson, you learned
- how to **solve** a linear programming problem by determining the constraints, finding the feasible region, determining the feasible solutions, and finding the optimum solution;
- how to **use** linear programming to answer questions about real-life situations;
- how to **find** the best solution when several conditions have to be met;
- how to **model** many business problems involving maximizing costs or minimizing expenses using linear programming.

Example

Sarah makes two types of quilts: a regular quilt, with 1 kg of down and 1.5 kg of feathers, and a deluxe quilt, with 1.5 kg of down and 1 kg of feathers. She has 14 kg of down and 13.5 kg of feathers, and she wants to make at least 3 deluxe quilts and 2 regular quilts. If the profit on a deluxe quilt is $120 and on a regular quilt is $100, how many of each quilt should she make to maximize profits?

Solution

Let d = number of deluxe quilts (horizontal axis)
Let r = number of regular quilts (vertical axis)

Determine the constraints.

a regular quilt, with 1 kg of down, and 1.5 kg of feathers, and a deluxe quilt, with 1.5 kg of down, ...
She has 14 kg of down.

Constraint on down:
$1r + 1.5d \leq 14$, or $r \leq -\frac{3}{2}d + 14$

A regular quilt, with ..., 1.5 kg of feathers, and a deluxe quilt, with ..., 1 kg of feathers. She has 13.5 kg of feathers.

Constraint on feathers:
$1.5r + 1d \leq 13.5$, or $r \leq -\frac{2}{3}d + 9$

She wants to make at least 3 deluxe quilts and 2 regular quilts.

Constraints on number to make: $d \geq 3$; $r \geq 2$

Find the feasible region.
Graph the solution of $d \geq 3$; $r \geq 2$.

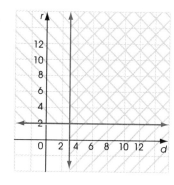

To this graph add $r \leq -\frac{3}{2}d + 14$.

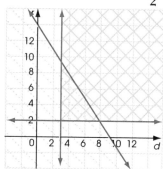

Next, add $r \leq -\frac{2}{3}d + 9$.

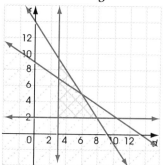

Finally, determine the solution of the complete system. Indicate corner points.

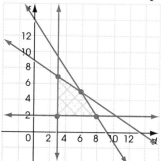

Determine feasible solutions.
Create a table of the corner points.

Deluxe Quilts	Regular Quilts	Profit $P = \$120d + \$100r$
3	2	$560
3	7	$1,060
6	5	$1,220
8	2	$1,160

Find the optimum solution.

If the profit on a deluxe quilt is $120, and the profit on a regular quilt is $100, how many of each quilt should she make to maximize profits?

The objective function is: $P = 120d + 100r$

From the table, it can be determined that a maximum profit of $1,220 is made if Sarah makes 6 deluxe quilts and 5 regular quilts.

Exercises

1. Find the maximum value of the quantity C, where $C = 3x + 5y$, given the constraints $x \geq 0$; $y \geq 0$; $x + 5y \leq 45$; $3x + 5y \leq 55$; and $3x + 2y \leq 40$.

2. Determine the feasible region, given the constraints $2x + 3y \leq 33$; $2x + y \geq 9$; $3x - y \geq 0$; and $x - 3y \leq -6$.
   (a) Determine both the minimum and the maximum values of the objective function $S = x + 4y$, and note the points at which they occur.

   (b) Determine both the minimum and the maximum values of the objective function $Q = x - 4y$, and note the points at which they occur.

   (c) Compare your answers for (a) and (b). What do you notice?

3. Determine the feasible region, given the constraints $x + 3y \geq 6$; $4x + 3y \leq 18$; $x - y \geq -3$; $x \geq 0$; and $y \geq 0$.
   The objective function is $R = 6x + 5y$.
   (a) Find the maximum value of the objective function, R, and the point at which it occurs.

   (b) Find the maximum value of the objective function, R, and the point at which it occurs, if both x and y must be natural numbers.

4. A sportswear company has a division that makes jeans and jean jackets. Each article requires time for cutting, sewing, and finishing, as shown in the table. The number of available minutes of labor for each process is also given.

	Jean Jackets	Jeans	Time Available
Cutting	4	2	440
Sewing	5	6	780
Finishing	6	4	680

If jeans sell for $45, and jean jackets sell for $65, how many of each article should be made each day to maximize earnings?

5. Terry spends his summers golfing and cycling. Depending on the weather, he engages in the two sports between 15 and 30 hours per week. He spends at least twice as much time on the golf course as on his bicycle, but he cycles at least 3 hours per week. If golf burns 300 calories an hour and cycling burns 420 calories an hour, find both the least number and greatest number of calories he will burn in a week.

6. Luxury Leather makes loveseats and sofas. Each week, they can produce 18 frames, for either loveseats or sofas. Leather costs $1,000 per sofa and $750 per loveseat. The weekly budget for leather is $14,000. Assembly time is 6 hours per sofa and 4 hours per loveseat. The total time available for assembly is 78 hours per week. They want to produce at least 6 loveseats each week. How many of each should they produce to maximize earnings if the earning:
   (a) per loveseat is $550 and per sofa is $800?

   (b) per loveseat is $450 and per sofa is $700?

   (c) for both loveseats and sofas is $500?

7. Nick has space for 48 bikes in his Bicycle Shop. His cost for an adult bike is $210, and the cost for a child bike is $90. He can spend a maximum of $7,350. He knows he will sell no more than twice as many adult bikes as child bikes. If the profit on an adult bike is $75, and the profit on a child bike is $45, how many of each bike should he buy to maximize his earnings?

8. Samara and her friend want to buy tickets to several concerts this year. They have a choice of two packages, broken down as follows:

	Jazz	Pop	Rock	Package Price
Package A	3 tickets	2 tickets	1 ticket	$114
Package B	1 ticket	6 tickets	1 ticket	$136

They want 20 jazz tickets, 24 pop tickets, and 8 rock tickets. Find the number of packages they should buy of each series to minimize their cost.

9 FUNCTIONS

9.1 Functions and Operations

You have seen that two functions, $f(x)$ and $g(x)$, can be added, subtracted, multiplied, or composed to produce a third function.

- Adding $f(x)$ and $g(x)$ results in a new function called $(f + g)(x)$.
- Subtracting $f(x)$ and $g(x)$ results in a new function called $(f - g)(x)$.
- Multiplying $f(x)$ and $g(x)$ results in a new function called $(f \bullet g)(x)$.
- $f(x)$ and $g(x)$ can be composed in two ways: $f(g(x))$ and $g(f(x))$

The value of a function for a given x-value is the vertical height of the point, or the second coordinate of the point $(x, f(x))$. The vertical heights can be used to perform operations on functions.

Example 1
If $f(x) = 4x - 3$ and $g(x) = x + 2$, find $(f \bullet g)(3)$. Check your answer.

Solution
Method 1
Evaluate $f(3)$ and $g(3)$ separately and multiply the results.

$$\begin{aligned}
(f \bullet g)(3) &= f(3) \bullet g(3) \\
&= [4(3) - 3][3 + 2] \\
&= (12 - 3)(5) \\
&= 9(5) \\
&= 45
\end{aligned}$$

Method 2
Find an expression for $f(x) \bullet g(x)$ and evaluate for 3.

$$\begin{aligned}
(f \bullet g)(x) &= f(x) \bullet g(x) \\
&= (4x - 3)(x + 2) \\
&= 4x^2 - 3x + 8x - 6 \\
&= 4x^2 + 5x - 6
\end{aligned}$$

$$\begin{aligned}
4x^2 + 5x - 6 &= 4(3)^2 + 5(3) - 6 \\
&= 4(9) + 15 - 6 \\
&= 36 + 15 - 6 \\
&= 51 - 6 \\
&= 45
\end{aligned}$$

Example 2

If $f(x) = 2x + 1$ and $g(x) = 3x^2$, find an expression for $g(f(x))$.

Solution

$$g\big(f(x)\big) = g(2x+1)$$
$$= 3(2x+1)^2$$
$$= 3\big(4x^2 + 4x + 1\big)$$
$$= 12x^2 + 12x + 3$$

Substitute $2x + 1$ for $f(x)$.

Find $g(2x + 1)$.

Exercises

1. Describe how the graphs of two functions, $f(x)$ and $g(x)$, can be added.

2. Describe how the graphs of two functions, $f(x)$ and $g(x)$, can be multiplied.

3. If $f(x) = 3x + 2$ and $g(x) = 6x$, find:
   (a) $(f - g)(x)$

   (b) $(f + g)(x)$

   (c) $(f \bullet g)(x)$

   (d) $(g - f)(x)$

4. If $f(x) = x + 4$, $g(x) = x^2$, and $h(x) = 5x$, find:
   (a) $(h \bullet f)(x)$

   (b) $g(x) - (f \bullet h)(x)$

   (c) $(f \bullet h)(x) - g(x)$

   (d) $(f \bullet g)(x) + h(x)$

   (e) $(g \bullet h)(x) - (g \bullet f)(x)$

5. If $f(x) = 4x + 1$, $g(x) = x^2$, and $h(x) = 3x$, find:

(a) $h(f(x))$

(b) $f(h(x))$

(c) $g(f(x))$

(d) $f(g(x))$

(e) $h(g(x)) - f(g(x))$

(f) $f(f(x))$

(g) $h(h(x))$

(h) $g(h(x)) + h(g(x))$

6. Sketch the graphs of $f(x) = 2x - 2$, $g(x) = x + 1$, and $(f \bullet g)(x)$.

7. Sketch the graphs of $f(x) = 3x$, $g(x) = x - 3$, and $(f - g)(x)$.

8. Sketch the graphs of $f(x) = x + 5$, $g(x) = x^2$, and $g(f(x))$.

9. The length of a rectangle is 5 units more than its width. If the original width is w, find an expression for the increase in area if each dimension is increased by 8.

10. The price of a CD player is reduced by 20% before a sales tax of 7% is added. Show how this is a composition of two functions, and find an expression for the final cost if the original cost of the CD player is C.

11. A stone is thrown into the air with a velocity given by $v(t) = 30 - 9.8t$. The kinetic energy is given by $K(v) = 0.4v^2$. Determine the kinetic energy as a function of the time, t.

12. The graphs of $f(x)$, $g(x)$, and $(f + g)(x)$ are shown. If $f(x) = x + 2$ and $g(x) = -2x - 1$, identify which graph is which.

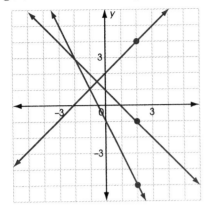

13. The table on the back of a pattern package can be used to determine the number of yards of fabric needed to make a rabbit costume for a child.

PATTERN 9810　　　　**Simplicity**

Costumes have front zipper, long raglan sleeves, elastic sleeve and leg casings, hood. Fabrics: Fleece or suede

BODY MEASUREMENTS

Chest (in.)	21	22	23	25	26	27	$28\frac{1}{2}$	29	30
Pattern Size	2	3	4	6	7	8	10	11	12

YARDAGE NEEDED

Pattern Size	2-4	3-8	12-14
Yards	$2\frac{5}{8}$	$3\frac{3}{8}$	$3\frac{3}{4}$

(a) How many yards of fabric are needed if the child's chest measures 29 inches?

(b) In this exercise, one number is a function of a second number that depends, in turn, on a third number. Explain this dependence.

14. Marnie combined $f(x)$ and $g(x)$ as shown. Find her error and write a correct solution.

Problem
If $f(x) = x^2 + 2x + 1$ and $g(x) = 5x$, find $f(g(3))$.
Solution

$$g(3) = 5(3)$$
$$= 15$$
$$f(g(3)) = 15(x^2 + 2x + 1)$$
$$= 15x^2 + 30x + 15$$

15. Create a question involving combinations of operations on functions, and then solve it. Exchange questions with a partner for checking.

Inverse Functions

In this lesson, you learned to find the inverse of a function.

- To find the inverse of a function algebraically:
  1. Change the function notation to y.
  2. Interchange x and y.
  3. Solve the resulting equation for y.

- To find the inverse of a function graphically:
  1. Select points from the function, interchange the coordinates, and plot these new points.
  2. Sketch the graph of the inverse by reflecting the function over the line $y = x$. This graph should go through the plotted points.

- To determine if the inverse of a function will also be a function, use the Horizontal Line Test on the original function.

- To determine if functions $f(x)$ and $g(x)$ are inverse functions, find $f(g(x))$ and $g(f(x))$. If $f(g(x)) = g(f(x)) = x$, then $f(x)$ and $g(x)$ are inverse functions.

Example 1

Find the inverse of $f(x) = -4x^3 + 2$.

Solution

$$f(x) = -4x^3 + 2 \quad \text{Change the function}$$
$$y = -4x^3 + 2 \quad \text{notation to } y.$$
$$x = -4y^3 + 2 \quad \text{Interchange } x \text{ and } y.$$
$$x - 2 = -4y^3 \quad \text{Solve the equation for } y.$$
$$\frac{-x + 2}{4} = y^3$$
$$\sqrt[3]{\frac{-x + 2}{4}} = y$$
$$f^{-1}(x) = \sqrt[3]{\frac{-x + 2}{4}}$$

Example 2

Sketch the inverse of the figure.

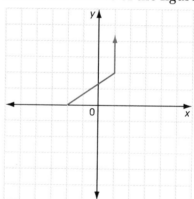

Solution

$(-2, 0)$, $(1, 2)$, $(1, 3)$, and $(1, 4)$ lie on the original graph, so $(0, -2)$, $(2, 1)$, $(3, 1)$, and $(4, 1)$ will lie on the inverse. Once these points are plotted, reflect the figure over the line $y = x$.

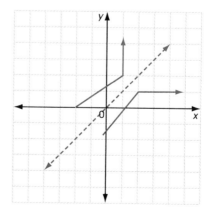

Exercises

1. If $f(x) = \frac{1}{3}x^3 - 1$, describe the steps needed to find $f^{-1}(x)$.

2. Explain how the Horizontal Line Test helps you to determine if the inverse of a function will be a function.

3. Given $g(x) = 2x^2 + 1$, explain why the notation $g^{-1}(x)$ cannot be used to denote the inverse of $g(x)$.

4. Represent the inverse of $f(x) = -3x + 6$ in two different ways. How could you show that these representations illustrate the same inverse?

5. Find $f^{-1}(x)$ for the following functions.
   (a) $f(x) = \{(-1, 2), (3, 4), (0, -7), (11, -6)\}$ (b) $f(x) = 4x - 2$

   (c) $f(x) = \frac{1}{x} + 4$ (d) $f(x) = x^3 + 2$

6. Sketch the inverse of each relation.
   (a)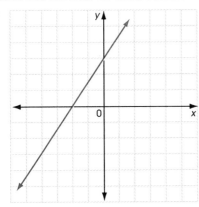

   (b)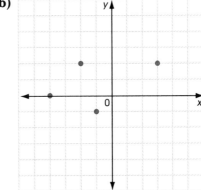

(c)

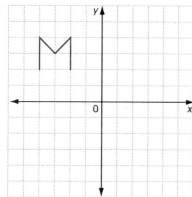

(d)

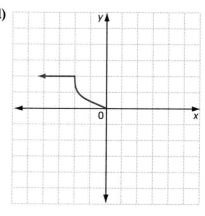

7. A friend is having trouble finding the inverse of $g(x) = \dfrac{3x}{x-2}$. He has done the following work:

$$g(x) = \frac{3x}{x-2}$$
$$y = \frac{3x}{x-2}$$
$$x = \frac{3y}{y-2}$$

Complete the remaining steps to find $g^{-1}(x)$.

8. State the inverse of each relation.
   (a) $y = -12$ (b) $y = (x+3)^2 - 1$ (c) $\{(3, -2), (2, 3), (2, 0)\}$

9. Explain why the inverses from Problem 8 are not functions.

10. If two functions $f(x)$ and $g(x)$ are inverse functions, then $f(g(x)) = g(f(x)) = x$.

   Use this to show that $h(x) = \dfrac{1}{3}x - 4$ and $k(x) = 3x + 12$ are inverse functions.

11. Which of the following functions have inverses that are *not* functions? Explain.
   (a)

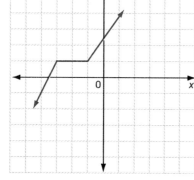

   (b)

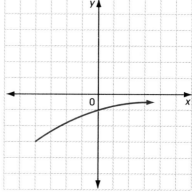

(c)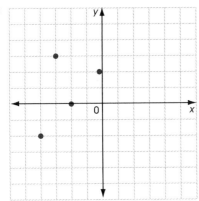

(d)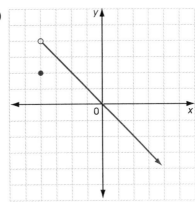

12. Given the graph of $f(x)$, sketch $f^{-1}(x)$.

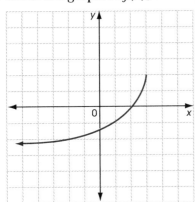

(a) State the domain and range of $f(x)$ and $f^{-1}(x)$.

(b) Compare the domain of $f(x)$ with the range of $f^{-1}(x)$ and the range of $f(x)$ with the domain of $f^{-1}(x)$. Explain what has occurred and why.

13. Find $f^{-1}(x)$ for each function below, and evaluate $f^{-1}(0)$ and $f^{-1}(-2)$.

(a) $f(x) = \frac{1}{2}x - 3$ **(b)** $f(x) = \frac{4}{x+1}$ **(c)** $f(x) = -x^3 + 8$

(d) $f(x) = \{(0, -2), (-2, 4), (3, 0)\}$ **(e)** $f(x) = \frac{2}{3}x - \frac{4}{3}$

14. Explain how you could evaluate $f^{-1}(0)$ for $f(x) = \frac{1}{2}x - 3$ without first finding $f^{-1}(x)$.

15. Sketch the graph of $h(x) = \sqrt{x+1}$. State the domain and range of $h(x)$. Then find $h^{-1}(x)$ algebraically and graphically. State the restrictions that need to be placed on the algebraic form so that it correctly represents the inverse.

16. Letitia stated that the function $f(x) = \frac{x}{x-1}$ does not have an inverse. Find $f^{-1}(x)$ and explain what might have led her to this incorrect conclusion.

9.3 Graphs of Quadratic Functions

In this lesson, you examined a number of characteristics and properties of the graphs of quadratic functions.

- Quadratic functions are second-degree polynomials with U-shaped graphs.
- The highest or lowest point on the graph is the **vertex**.
- The y-coordinate of the vertex is the **maximum** or **minimum value** of the function.
- The vertical line that passes through the vertex is called the **axis of symmetry** because it divides the parabola into two equal halves.
- An unrestricted quadratic function will have 1 y-intercept, and 0, 1, or 2 x-intercepts.
- The domain of a function is all possible x-values for which the function is defined. The range is all possible y-values for which the function is defined.

Example 1

Graph the quadratic function defined by the equation $y = 0.5x^2 - 2x$ to determine its characteristics.

Solution

Plot the graph using a graphing tool or a table of values.

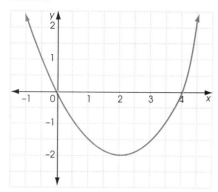

Characteristics
Direction of opening: up
Maximum/minimum value: minimum value of –2
Vertex: (2, –2)
Equation of the axis of symmetry: $x = 2$
x-intercepts: (0, 0) and (4, 0)
y-intercept: (0, 0)
Domain: $\{x \mid x \in \mathrm{R}\}$
Range: $\{y \mid y \geq -2, y \in \mathrm{R}\}$

Example 2

Find two numbers that have a sum of 52 and whose product is a maximum.

Solution

Step 1: Think about the problem.
Identify the unknowns.
　　　　Let x = the first number
　　　　Let $52 - x$ = the second number

Step 2: Make a plan.
Write an equation for the problem.
　　　$P = x(52 - x)$

Step 3: Solve the problem.
Plot the graph and find where the maximum occurs.

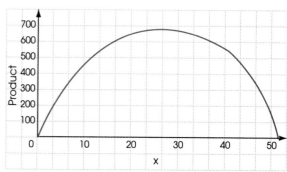

You may need to zoom in on parts of the graph to determine exact values. The vertex is at (26, 676). The two numbers are 26 and 26, and the maximum product is 676.

Exercises

1. How can you tell whether a graph has a maximum value or a minimum value?

2. Tell whether the graph opens up or down, and give the maximum/minimum value.

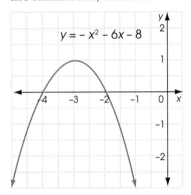

3. Use a graphing tool to plot the function $y = x^2 - 2x + 1$. Tell whether it opens up or down, and give the maximum/minimum value.

4. How does the vertex of a graph relate to each of the following?
   (a) maximum/minimum value

   (b) equation of the axis of symmetry

5. Use a graphing tool to find the coordinates of the vertex, the maximum/minimum value, and the equation of the axis of symmetry for the graph.

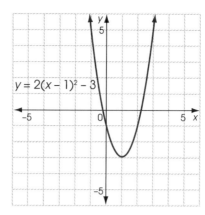

6. Identify the x- and y-intercepts from the graph.

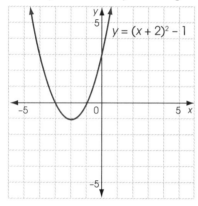

7. Plot the function $y = -x^2 - 1$. Identify the number of x- and y-intercepts and their coordinates.

8. Determine the domain and range of the graph in Problem 6.

9. Determine the domain and range for the restricted quadratic function.

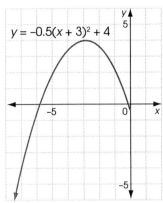

$y = -0.5(x + 3)^2 + 4$

10. Describe a real-life problem that might be represented by a quadratic function. Why might someone want to identify the vertex and the intercepts?

9.4 Completing the Square

You have learned these steps to complete the square:
1. Group the x^2- and x-terms together inside parentheses.
2. Factor out a, the coefficient of the x^2-term. Adjust the coefficient of the x-term.
3. Create a perfect square trinomial. Subtract the same value that was added for the trinomial.
4. Simplify, and express the trinomial as a squared binomial.

Example 1

Describe the graph of the quadratic function $y = -4x^2 - 9x + 11$ as compared to the graph of the basic quadratic function, $y = x^2$.

Solution

General form: $y = ax^2 + bx + c$
$$y = -4x^2 - 9x + 11$$

$a = -4$ Since a is negative, the parabola will open downward.

$|a| = 4$ Since $|a| > 1$, the parabola will be narrower than the basic quadratic function, $y = x^2$.

Therefore, the parabola of this quadratic function will open downward and will be narrower than the basic $y = x^2$ graph.

Example 2

Complete the square for the quadratic function

$y = \frac{1}{2}x^2 - 3x + 5$. Find the vertex and the

y-intercept, then use these two points to graph the parabola.

Solution

❶ $y = (\frac{1}{2}x^2 - 3x) + 5$

❷ $y = \frac{1}{2}(x^2 - 6x) + 5$

❸ $y = \frac{1}{2}(x^2 - 6x + 9) + 5 - \frac{1}{2}(9)$

❹ $y = \frac{1}{2}(x - 3)^2 + \frac{10}{2} - \frac{9}{2}$

$y = \frac{1}{2}(x - 3)^2 + \frac{1}{2}$

❶ Group the x^2- and x-terms together inside parentheses.
❷ Factor out a, the coefficient of the x^2-term. Adjust the coefficient of the x-term.
❸ Create a perfect square trinomial. Subtract the same value that was added for the trinomial.

❹ Simplify, and express the trinomial as a squared binomial.

vertex coordinates: $(3, \frac{1}{2})$

To find the y-intercept, set $x = 0$.

$y = \frac{1}{2}x^2 - 3x + 5$

$y = \frac{1}{2}(0)^2 - 3(0) + 5$ Substitute $x = 0$.

$y = 0 - 0 + 5$ Simplify.

$y = 5$

y-intercept: $(0, 5)$

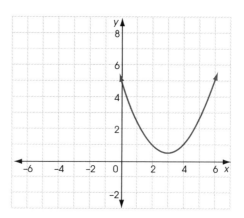

Exercises

1. Give a list of events or situations that might be described by a quadratic, parabolic graph.

2. Explain how you would recognize that an equation represents a quadratic function.

3. In standard form, $y = a(x - h)^2 + k$, what do the letters a, h, and k represent?

4. Explain:
   (a) how the value of a affects the direction of opening of a parabola.
   (b) how the value of a affects the shape of a parabola.

5. Is each statement true or false? Reword the false statements to make them true.
   (a) $y = -3x^2 + 8$
       This parabola opens downward and is narrower than the basic $y = x^2$ graph.
   (b) $y = \frac{1}{5}x^2 - 3x + 14$
       This parabola opens upward and is narrower than the basic $y = x^2$ graph.
   (c) $y = \frac{5}{4}(x - 7)^2 + 20$
       This parabola opens upward and is wider than the basic $y = x^2$ graph.
   (d) $y = -\frac{2}{3}(x + 1)^2 - 9$
       This parabola opens downward and is wider than the basic $y = x^2$ graph.

6. Select the equation of the parabola.
   (a)
   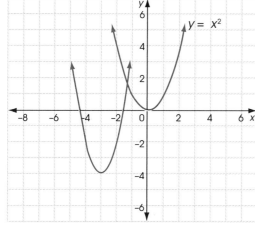
   (b)
   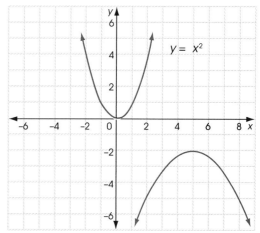

   (i) $y = -2(x + 3)^2 - 4$

   (ii) $y = -2(x - 3)^2 - 4$

   (iii) $y = 2(x - 3)^2 - 4$

   (iv) $y = 2(x + 3)^2 - 4$

   (i) $f(x) = \frac{1}{3}(x - 5)^2 - 2$

   (ii) $f(x) = -\frac{1}{3}(x - 5)^2 - 2$

   (iii) $f(x) = 3(x - 5)^2 - 2$

   (iv) $f(x) = -3(x - 5)^2 - 2$

(c)

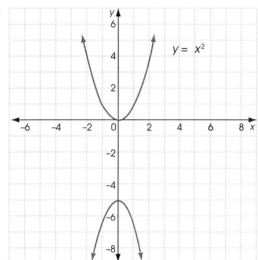

(d)

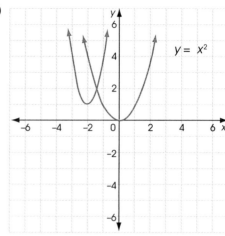

(i) $y = \dfrac{3}{2}x^2 - 5$

(ii) $y = -\dfrac{3}{2}x^2 - 5$

(iii) $y = \dfrac{2}{3}x^2 - 5$

(iv) $y = -\dfrac{2}{3}x^2 - 5$

(i) $y = 3(x+2)^2 + 1$

(ii) $y = 3(x+2)^2 - 1$

(iii) $y = 3(x-2)^2 + 1$

(iv) $y = 3(x-2)^2 - 1$

7. The summary lists the four steps involved in completing the square. Perform steps 1 and 2 for each quadratic equation.

(a) $y = -7x^2 - 14x + 9$

(b) $f(x) = -2.3x^2 + 9.2x - 5.4$

(c) $y = 4x^2 - 2x + 5$

(d) $y = 15x^2 - 5x + 2$

(e) $y = \dfrac{1}{3}x^2 + \dfrac{1}{5}x - \dfrac{4}{9}$

(f) $f(x) = \dfrac{3}{8}x^2 + \dfrac{1}{4}x + \dfrac{1}{2}$

8. An error was made in completing the square of each of these functions. State the line in which the error first occurs, then correct the error and redo the solution.

(a) $y = x^2 - 24x + 100$
$y = (x^2 - 24x) + 100$
$y = (x^2 - 24x + 144) + 100 - 144$
$y = (x - 12)^2 + 44$

(b) $y = 5x^2 - 20x + 4$
$y = 5(x^2 - 4x) + 4$
$y = 5(x^2 - 4x + 4) + 4 - 20$
$y = 5(x - 4)^2 - 16$

(c) $y = 12x^2 + 6x - 11$

$y = 12(x^2 - 2x) - 11$

$y = 12(x^2 - 2x + 1) - 11 - 12$

$y = 12(x - 1)^2 - 23$

(d) $f(x) = -4x^2 - 32x + 3$

$f(x) = -4(x^2 + 8x) + 3$

$f(x) = -4(x^2 + 8x + 64) + 3 + 256$

$f(x) = -4(x + 8)^2 + 259$

9. Rewrite each equation in the form $y = a(x - h)^2 + k$.

(a) $y = x^2 + 10x + 28$

(b) $f(x) = 2x^2 + 8x + 7$

(c) $y = 1 - 18x + 3x^2$

(d) $y = x^2 - 3x + 7$

10. Rewrite each equation to find the vertex and the y-intercept.

(a) $y = -2x^2 + 12x - 17$

(b) $y = \frac{1}{2}x^2 + 2x - 3$

(c) $f(x) = \frac{1}{4}x^2 - 2x - 6$

(d) $y = 6x - \frac{3}{2}x^2 - 31$

11. Use the formula $V_x = -\frac{b}{2a}$ to find the x-coordinate of the vertex. Use this value to find the y-coordinate. State the y-intercept.

(a) $y = 3x^2 - 6x + 1$

(b) $f(x) = -\frac{1}{4}x^2 - 2x + 6$

(c) $y = 4x - 1 + \frac{2}{3}x^2$

(d) $y = -\frac{2}{3}x^2 - \frac{8}{3}x + \frac{1}{2}$

12. Find the coordinates of the vertex and the y-intercept. Graph each quadratic function.

(a) $y = 10 - 10x + x^2$

(b) $f(x) = -3x^2 + 12x - 8$

(c) $y = -\frac{1}{2}x^2 - 3x - \frac{5}{2}$

(d) $y = \frac{3}{2}m^2 - 12m + 10$

Translations of Quadratic Functions

The standard form of a quadratic equation is $y = a(x - h)^2 + k$.

- Its vertex is at (h, k) and its axis of symmetry is $x = h$.
- If $a > 0$ then the parabola opens upward; if $a < 0$ then the parabola opens downward.
- Increasing the value of a makes the parabola narrower; decreasing the value of a makes the parabola wider.
- The constant h represents a horizontal translation by h units (right if h is positive, left if h is negative).
- The constant k represents a vertical translation by k units (up if k is positive, down if k is negative).
- Every point on a parabola has a mirror point that is its reflection in the axis of symmetry.

Example 1

Graph $f(x) = 3(x - 2)^2 + 5$.

Solution

The graph of $f(x) = 3(x - 2)^2 + 5$ is identical to the graph of $g(x) = 3(x - 2)^2$, except that it has been translated 5 units upward. The graph of $g(x) = 3(x - 2)^2$ is identical to the graph of $h(x) = 3x^2$, except that it has been translated 2 units to the right. Thus, to graph $f(x) = 3(x - 2)^2 + 5$, we can graph $h(x)$ and shift it 2 units to the right and 5 units up.

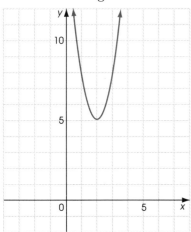

The vertex of the graph is the point $(2, 5)$, and the axis of symmetry is the line $x = 2$.

Example 2

For $y = -4(x + 3)^2 - 1$, state the direction in which the graph opens, the vertex, and the axis of symmetry.

Solution

Rewrite the function in $f(x) = a(x - h)^2 + k$ form.
$$f(x) = -4[x - (-3)]^2 + (-1)$$
$$\quad\quad\;\; \downarrow \quad\quad\;\; \downarrow \quad\quad\;\; \downarrow$$
$$\quad\quad\;\; a \quad\quad\;\; h \quad\quad\;\; k$$

Since $a = -4$, then $a < 0$ and the parabola opens downward.

The vertex is $(h, k) = (-3, -1)$

The axis of symmetry is the line $x = h$. In this case the line is $x = -3$.

Exercises

1. The function $f(x) = 2(x-1)^2 - 1$ is shown. Fill in the blanks to make the statement true.

   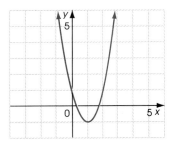

   (a) This is a _____ function.

   (b) The lowest point on the graph is $(1, -1)$. This is called the _____.

   (c) The vertical line $x = 1$ divides the parabola into two halves. This line is called the _____.

2. Is it necessary to calculate a new table of values to sketch the graphs of parabolas congruent to $y = x^2$? Explain.

3. Complete the table of values for each set of functions. Then graph each set on the same grid.

   (a)

x	$f(x) = x^2$	$g(x) = 2x^2$	$h(x) = \dfrac{1}{2}x^2$
-3			
-2			
-1			
0			
1			
2			
3			

   (b)

x	$f(x) = -x^2$	$g(x) = -\dfrac{1}{4}x^2$	$h(x) = -4x^2$
-3			
-2			
-1			
0			
1			
2			
3			

4. The function $f(x) = 2(x + 1)^2 + 6$ is written in the form $f(x) = a(x - h)^2 + k$. Is $h = -1$ or is $h = 1$? Explain.

5. Make a table of values to graph $f(x)$. Graph the other two functions by finding the correct translation.

(a) $f(x) = 2x^2$, $g(x) = 2x^2 + 3$, $h(x) = 2x^2 - 1$

(b) $f(x) = 3x^2$, $g(x) = 3(x + 2)^2$, $h(x) = 3(x - 3)^2$

6. Make a table of values to graph $f(x)$. Then use a series of translations to graph $g(x)$ on the same grid.

(a) $f(x) = -3x^2$, $g(x) = -3(x - 2)^2 - 1$

(b) $f(x) = -\dfrac{1}{2}x^2$, $g(x) = -\dfrac{1}{2}(x + 1)^2 + 2$

7. For each function, state the axis of symmetry and the mirror point of the given point.

(a) $y = 2(x - 3)^2$, $(0, 18)$

(b) $y = 4x^2 + 3$, $(-2, 19)$

(c) $y = \dfrac{1}{2}(x + 2)^2 - 3$, $(4, 15)$

(d) $y = (x - 5)^2 + 6$, $(-1, 42)$

8. For each set of graphs, find the equation of the translated graphs given $f(x)$.

(a) $f(x) = -2x^2$

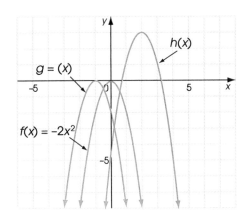

(b) $f(x) = \frac{1}{2}x^2$

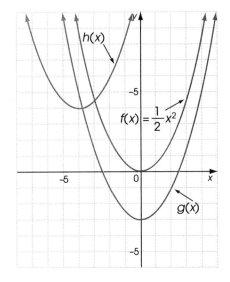

(c) $f(x) = 3x^2$

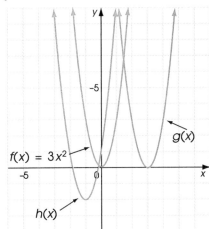

9. Determine the vertex and the axis of symmetry of the graph of the function. Then state if the graph will open upward or downward. Do not graph the function.

(a) $y = (x-1)^2 + 2$

(b) $y = -2(x-2)^2 - 1$

(c) $y = 2(x+3)^2 - 4$

10 QUADRATIC EQUATIONS

10.1 Solving Quadratics by Graphing

You have seen how to solve a quadratic equation by graphing the equation and finding the *x*-intercepts. These intercepts represent the **zeros** of the function, since the value of the function is zero when *x* has these values. They are also called the **roots** of the equation.

Example

Estimate the roots of $2x^2 - 12x - 6 = 0$.

Solution

Graph the function with a graphing utility or by creating a table of values.

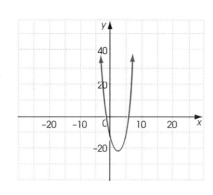

You may wish to verify the coordinates of the graph vertex by completing the square of the function.

To change from one general form of the equation, $ax^2 - 6x + c = 0$, to the standard form, $a(x - h)^2 + k = f(x)$, follow these steps:

$$f(x) = 2x^2 - 12x - 6$$

Write the function.

$$f(x) = 2(x^2 - 6x - 3)$$

Factor out the *a* term, 2.

$$f(x) = 2(x^2 - 6x + 9 - 12)$$

Complete the square. Since the first part of the expression is $x^2 - 6x$, the square root must be $(x - 3)$ and the square must be $x^2 - 6x + 9$.

$$f(x) = 2(x - 3)^2 - 24$$

Factor the square and separate *k*.

The value of *a* is 2, so the graph is a transformation of $y = 2x^2$.

Since *a* is positive, the graph opens up, not down.

The values for *h* and *k* are 3 and –24, so the vertex is at (h, k) or $(3, -24)$.

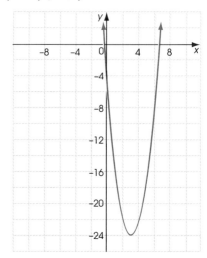

The *x*-intercepts are at about –0.5 and 6.5, so these are the roots of the equation (and the zeros of the function).

Check by substituting these values for x into the equation.

If the estimates are reasonable, the value of
$2x^2 - 12x - 6$ should be near 0.

For $x = -0.5$,

L.S.	R.S.
$2x^2 - 12x - 6$	0
$= 2(-0.5)^2 - 12(-0.5) - 6$	
$= 0.5 + 6 - 6$	
$= 0.5$	

For $x = 6.5$,

L.S.	R.S.
$2x^2 - 12x - 6$	0
$= 2(6.5)^2 - 12(6.5) - 6$	
$= 84.5 - 78 - 6$	
$= 0.5$	

Both equations simplify to values near 0, so the estimates are reasonable.

Exercises

1. Match each term from column A with a term from column B. Then give an example that illustrates each pair of terms.

A	B
equation	x-intercept
function	root
graph	zero

2. Sketch a graph to identify a function that fits each description.
   (a) no x-intercepts

   (b) one x-intercept

   (c) two x-intercepts

3. Is it possible for a quadratic function to have more than two roots? Explain.

4. Verify to determine if each root matches the given equation. Identify any roots that are close estimates, as well as those that are exact.
   (a) $x = 5$, $x^2 + 3x = 40$ **(b)** $x = \frac{2}{3}$, $3x^2 + x = 2$

   (c) $x = 2.5$, $6x^2 + 17x + 5 = 0$ **(d)** $x = 0.75$, $3(4x^2 - 1) = -5x$

(e) $x = 2.645$, $2x^2 = 5x + 1$

(f) $x = -1.41$, $x^2 = 4x + 7$

(g) $x = 1.73$, $3x^2 - 4x = 1$

5. Express each equation in the form $ax^2 + bx + c = 0$.
Use a graphing utility to graph each equation and then find or estimate the roots.
   (a) $2x^2 - 6x = 0$ **(b)** $x^2 + 5x = 14$ **(c)** $x^2 + 5 = 13x$

   (d) $x(2x + 5) = 11$ **(e)** $3x(2x + 5) = -7x - 8$ **(f)** $(x - 3)(x + 4) = 3(x + 1)$

6. Use a graphing utility to graph each function and then find or estimate the zeros where they exist.
   (a) $f(x) = (x + 5)^2 - 4$ **(b)** $f(x) = 2(x - 3)^2 - 2$ **(c)** $f(x) = -0.5(x + 4)^2 + 6$

   (d) $f(x) = 3x^2 + 12x - 36$ **(e)** $f(x) = -0.5x^2 + 3x - 5$ **(f)** $f(x) = 3x^2 + 5x - 7$

   (g) $f(x) = 3x^2 + 5x + 4$

7. Find and correct any errors in Leticia's problem solution.

Problem:

Are there three consecutive odd integers whose squares have a sum of 155?

Solution:

If x is the first integer, then $x + 2$ and $x + 4$ are the other two consecutive integers.

$$x^2 + (x + 2)^2 + (x + 4)^2 = 155$$
$$x^2 + (x^2 + 4) + (x^2 + 16) = 155$$
$$3x^2 + 20 = 155$$
$$3x^2 - 135 = 0$$
$$f(x) = 3x^2 - 135$$

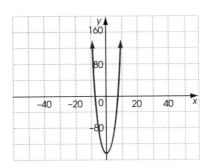

The x-intercepts are located at about 7 and –7.

This means that one of the integers is either 7 or –7.

The integers are 5, 7, and 9 or –5, –7, and –9.

8. The sum of the squares of two consecutive numbers is 421. Find the numbers.

9. Find two consecutive numbers whose product is 552.

10. Create a quadratic equation. Use a graphing utility to graph your equation and find (or estimate) its roots.

Use your equation and graph to create an error analysis problem similar to the one in Problem 7. Exchange problems with a classmate and then find and correct the errors in one another's problems.

11. Talk with a partner about the advantages and disadvantages of using graphs to solve quadratic equations. Write a brief summary based on your observations.

Solving Quadratics by Factoring

You have seen how to determine the zeros of a quadratic function by factoring.
- Set the function equal to 0.
- Factor the function.
- Set each factor equal to zero.
- Solve for x to determine the zeros.

A zero is a value of x that makes a function equal to zero. This value is the x-coordinate of a point where the graphed function touches the x-axis. (This point is also called an x-intercept.) A quadratic function may have 0, 1, or 2 zeros. When the function is expressed as an equation, the zeros are called **roots** or **solutions**.

Example 1

Solve the equation, $3x^2 - 11x = 14$, by factoring. Verify the roots.

Solution

$$3x^2 - 11x = 14 \qquad \text{Change the equation to the } = 0 \text{ form.}$$
$$3x^2 - 11x - 14 = 0 \qquad \text{Factor the trinomial.}$$
$$(3x - 14)(x + 1) = 0 \qquad \text{Set each factor equal to 0.}$$
$$3x - 14 = 0 \quad x + 1 = 0 \qquad \text{Solve for } x.$$
$$3x = 14 \qquad x = -1$$
$$x = \frac{14}{3}$$

Verify each root by substituting the value into the equation $3x^2 - 11x = 14$.
If the root is correct, the left side should equal the right side.

For $x = \frac{14}{3}$,

L.S.	R.S.
$3\left(\frac{14}{3}\right)^2 - 11\left(\frac{14}{3}\right)$	14
$= \frac{196}{3} - \frac{154}{3}$	
$= \frac{42}{3}$	
$= 14$	

L.S. = R.S.

For $x = -1$,

L.S.	R.S.
$3(-1)^2 - 11(-1)$	14
$= 3 + 11$	
$= 14$	

L.S. = R.S.

Example 2

The zeros of a quadratic function are $\frac{5}{4}$ and $-\frac{2}{3}$.

What is the function?

Solution

You know that the function is equal to 0 if $x = \frac{5}{4}$ or $x = -\frac{2}{3}$.

$$\left(x - \frac{5}{4}\right) \qquad \text{Select a factor that has a value of 0 when } x \text{ is } \frac{5}{4}.$$

$$\left(x + \frac{2}{3}\right) \qquad \text{Select another factor that has a value of 0 when } x \text{ is } -\frac{2}{3}.$$

$$f(x) = \left(x - \frac{5}{4}\right)\left(x + \frac{2}{3}\right) \qquad \text{Write the function using the two factors you found.}$$

$$f(x) = x^2 - \frac{5}{4}x + \frac{2}{3}x - \frac{10}{12} \qquad \text{Find the product of the two factors.}$$

$$f(x) = x^2 - \frac{7}{12}x - \frac{5}{6} \qquad \text{Simplify.}$$

To find other functions with the same zeros, multiply all the terms in the function by the same number. For example, multiply by 12 to get the function $12x^2 - 7x - 10 = 0$.

Exercises

1. Define each term.
    (a) quadratic equation

    (b) quadratic function

    (c) graph of a quadratic

    (d) verify

    (e) root

    (f) zero

    (g) x-intercept

2. Verify to determine if the numbers are roots of the given quadratic equations.

	Root	Quadratic Equation
(a)	-5	$2x^2 + 3x - 5 = 0$
(b)	$\frac{1}{2}$	$4x^2 + 4x - 5 = 0$
(c)	$-\frac{1}{2}$	$2x^2 - 9x - 5 = 0$
(d)	2.4	$12x^2 - 7x - 12 = 0$
(e)	-3	$5x^2 + 3x - 15 = 0$

3. Find the roots.
    (a) $3x^2 - 15x = 0$

    (b) $x^2 - 3x = 10$

    (c) $x^2 + 28 = 11x$

    (d) $x(2x - 4) = 6$

    (e) $3x(3x - 2) = 6x + 5$

    (f) $(x - 2)(x + 1) = 4(x + 1)$

4. Find the zeros.

    (a) $f(x) = x^2 + 7x + 12$

    (b) $f(x) = 2x^2 - 3x - 2$

    (c) $f(x) = 3x^2 + 6x - 24$

    (d) $f(x) = 10x^2 + 19x + 9$

    (e) $f(x) = 2 + 5x + 2x^2$

    (f) $f(x) = 4x^2 + 12x + 5$

5. Write the function with the given roots.

    (a) $3, 5$

    (b) $-2, \dfrac{1}{2}$

    (c) $\dfrac{3}{5}, \dfrac{3}{5}$

    (d) $-\dfrac{3}{4}, -\dfrac{4}{5}$

    (e) $0, -\dfrac{6}{7}$

    (f) $\dfrac{5}{2}, -\dfrac{5}{2}$

6. Check the solution to this problem and correct any errors.

Problem

Find two numbers that differ by 16 and have a product of 46.25.

Solution

1. Let one number be x and the other number be $x + 16$.

2. $\quad\quad x(x + 16) = 46.25$

3. $\quad\quad x^2 + 16x = 46.25$

4. $\quad x^2 + 16x - 46.25 = 0$

5. Multiply both sides by 4 to eliminate the decimal.

6. $\quad 4x^2 + 64x - 185 = 0$

7. $\quad (2x - 5)(x + 37) = 0$

8. $\quad\quad$ If $2x - 5 = 0$,

9. $\quad\quad\quad 2x = 5$

10. $\quad\quad\quad x = \dfrac{5}{2} = 2.5$

11. $\quad\quad$ If $x + 37 = 0$,

12. $\quad\quad\quad$ then $x = -37$.

13. If x is 2.5, then $x + 16$ must be 18.5.

14. If x is -37, then $x + 16$ must be -21.

15. There are two possible solutions to the problem: 2.5 and 18.5 or -37 and -21.

7. Find two consecutive integers whose product is 756.

8. The area of a rectangle in square feet is $x^2 + 13x + 36$. How much longer than the width is the length of the rectangle?

9. The volume of a rectangular box with a square base is $27x^2 - 90x + 75$. If the box is 3 units high, find the perimeter of the base.

10. Some word problems involve restrictions that eliminate some zeros as possible answers. Create a problem of this type so one of the zeros is not a reasonable solution.

10.3 Complex Numbers

You have learned what imaginary and complex numbers are and how to perform basic operations on them.

- An imaginary number is a number with a negative square.
- $i^1 = i$, $i^2 = -1$, $i^3 = -i$, and $i^4 = 1$. For i with a degree greater than 4, if n is a natural number that has a remainder of r when divided by 4, then $i^n = i^r$.
- A complex number is written in the form $a + bi$, where a is the real part, bi is the imaginary part, a and b are real numbers, and $i = \sqrt{-1}$.
- Always express the square roots of negative numbers in terms of i before adding, subtracting, multiplying, or dividing them.
- To add (or subtract) two complex numbers, add (or subtract) their real parts and add (or subtract) their imaginary parts.
- Complex numbers can be multiplied as if they were binomials using the FOIL method.
- To divide by a complex number, use complex conjugates. The expressions $a + bi$ and $a - bi$ are complex conjugates.

Example 1

Express $\sqrt{-75}$ in terms of i.

Solution

$$\sqrt{-75} = \sqrt{25}\sqrt{3}\sqrt{-1}$$
$$= 5\sqrt{3}i$$

Example 2

Simplify i^{22}.

Solution

Recall that $i^4 = 1$.

$$i^{22} = (i^4)(i^4)(i^4)(i^4)(i^4)(i^2)$$
$$= (1)(1)(1)(1)(1)(i^2)$$
$$= i^2$$
$$= -1$$

Example 3

Perform the operation and express the answer in $a + bi$ form.

$$(3 + 4i) - (4 - 7i)$$

Solution

$$(3 + 4i) - (4 - 7i) = (3 - 4) + [4 - (-7)]i$$
$$= -1 + 11i$$

Example 4

Express $\dfrac{2}{3 + i}$ in $a + bi$ form.

Solution

Multiply by $\dfrac{3 - i}{3 - i}$ to eliminate i from the denominator.

$$\frac{2}{3 + i} = \frac{2}{3 + i} \cdot \frac{3 - i}{3 - i}$$
$$= \frac{6 - 2i}{9 - 3i + 3i - i^2}$$
$$= \frac{6 - 2i}{9 - (-1)}$$
$$= \frac{6}{10} - \frac{2i}{10}$$
$$= \frac{3}{5} - \frac{1}{5}i$$

Example 5

Solve the equation and express the solution in $a + bi$ form.

$$3x^2 - 4x + 2 = 0$$

Solution

$$x = \frac{4 \pm \sqrt{(4)^2 - 4(3)(2)}}{2(3)}$$
$$x = \frac{4 \pm \sqrt{-8}}{6}$$
$$x = \frac{4 \pm 2\sqrt{2}i}{6}$$
$$x = \frac{2}{3} + \frac{\sqrt{2}}{3}i, \ \frac{2}{3} - \frac{\sqrt{2}}{3}i$$

Exercises

1. State when a complex number is the solution for an equation. Give two equations that have complex number solutions.

2. Which of the following operations will always result in a real number?
   (a) $(a + bi) + (a - bi)$

   (b) $(a + bi) - (a - bi)$

   (c) $(a + bi)(a - bi)$

3. Explain what a complex conjugate is. Then explain how and why it is used in the division of complex numbers.

4. Tell whether each statement is true or false.
   (a) $\sqrt{6}i = \sqrt{-6}$ (b) $\sqrt{8}i = \sqrt{8i}$ (c) $\sqrt{-25} = -\sqrt{25}$

   (d) $i^2 = -1$ (e) $\sqrt{-3}\sqrt{-2} = \sqrt{6}$ (f) $\sqrt{-36} + \sqrt{-25} = \sqrt{-61}$

5. Express each number in terms of i.
   (a) $\sqrt{-9}$ (b) $\sqrt{-11}$ c) $\sqrt{-24}$

   (d) $-\sqrt{-72}$ (e) $5\sqrt{-81}$ (f) $\sqrt{\dfrac{-25}{9}}$

   (g) $\sqrt{-9}\sqrt{-100}$ (h) $-\dfrac{\sqrt{-25}}{\sqrt{-64}}$

6. Solve each equation. Write all solutions in bi or $a + bi$ form.
   (a) $a^2 = -25$ (b) $2x^2 = -25$

   (c) $x^2 - 2x + 13 = 0$ (d) $2x^2 + 4x = -5$

7. Perform the operation and express each answer in $a + bi$ form.

(a) $(5 + 4i) + (7 - 12i)$

(b) $(-6 - 40i) - (-8 + 28i)$

(c) $(-8 + \sqrt{-8}) + (6 - \sqrt{-32})$

(d) $2i(64 + 9i)$

(e) $(2 - 7i)(-3 + 4i)$

(f) $(5 - \sqrt{-27})(-6 + \sqrt{-12})$

8. Write each expression in $a + bi$ form.

(a) $\dfrac{6}{2 + i}$

(b) $\dfrac{4 + i}{4 - i}$

(c) $\dfrac{\sqrt{3} + \sqrt{-4}}{\sqrt{3} - \sqrt{-4}}$

(d) $\dfrac{-2}{5i^3}$

9. Simplify each power of i.

(a) i^{34}

(b) i^{87}

(c) i^{65}

(d) i^{48}

10. In a series electric circuit, the total impedance is determined by adding up all of the individual impedances of the loads. If the total impedance of a series AC circuit with three loads is $(17 + 23i)$ ohms, and two loads are each $(4 + 5i)$ ohms, what is the impedance of the third load?

11. If $(a - 3i)(4 + bi) = 31 - 5i$, and a and b are integers, find a and b.

12. Find a complex number that results in 41 when multiplied by its complex conjugate.

13. Write a problem involving an AC circuit in which the impedance and electric potential are complex numbers and the current must be determined using Ohm's Law. Then solve the problem.

You have seen how to use the **quadratic formula** to find roots for **equations that cannot be factored**. The quadratic formula is derived by reorganizing the equation $ax^2 + bx + c = 0$ so x is **isolated** on one side.

The quadratic formula states that the two roots of $ax^2 + bx + c = 0$, where

$a \neq 0$, are $\quad x = \dfrac{-b + \sqrt{b^2 - 4ac}}{2a}$ and $\quad x = \dfrac{-b - \sqrt{b^2 - 4ac}}{2a}$.

The part of the quadratic formula located under the square root sign, $b^2 - 4ac$, is called the **discriminant**. When you simplify the discriminant using values from a given equation, you can **predict** how many **roots** the equation will have.

- If $b^2 - 4ac > 0$, the equation has two roots.
- If $b^2 - 4ac = 0$, the equation has one root.
- If $b^2 - 4ac < 0$, the equation has no roots.

Example

Solve $3x^2 - 5x - 14 = 0$. Verify the roots.

Solution

Step 1: Use the quadratic formula to find values for x.

$3x^2 - 5x - 14 = 0$	Make sure the equation is in the form $ax^2 + bx + c = 0$.
$a = 3,\ b = -5,\ c = -14$	Identify the values of a, b, and c.
$b^2 - 4ac = (-5)^2 - 4(3)(-14)$	Find the value of the discriminant. (Since the discriminant is 193, which is > 0, this equation must have two roots.)
$x = \dfrac{-b \pm \sqrt{b^2 - 4ac}}{2a}$	
$= \dfrac{5 \pm \sqrt{193}}{6}$	Substitute values for the discriminant, $-b$, and a in the formula
$x = \dfrac{5 + \sqrt{193}}{6} \qquad x = \dfrac{5 - \sqrt{193}}{6}$	Simplify to find values for x.
$x \approx 3.1 \qquad\qquad x \approx -1.5$	

Step 2: Verify values for x.

If the roots are correct, the value of the equation will be 0 when each root is substituted for x. (There may be a slight variation from 0 due to approximate numbers used in the calculations.)

For $x = \dfrac{5 + \sqrt{193}}{6}$

L.S.	R.S.
$3x^2 - 5x - 14$	0
$= 3\left(\dfrac{5 + \sqrt{193}}{6}\right)^2 - 5\left(\dfrac{5 + \sqrt{193}}{6}\right) - 14$	
$\approx 3(9.91) - 5(3.15) - 14$	
$= 29.73 - 15.7 - 14$	
$= 0.03$	

For $x = \dfrac{5 - \sqrt{193}}{6}$

L.S.	R.S.
$3x^2 - 5x - 14$	0
$= 3\left(\dfrac{5 - \sqrt{193}}{6}\right)^2 - 5\left(\dfrac{5 - \sqrt{193}}{6}\right) - 14$	
$\approx 3(2.20) - 5(-1.48) - 14$	
$= 6.6 + 7.4 - 14$	
$= 0$	

One equation is equal to 0, and the other is extremely close, so the roots appear to be correct.

Exercises

1. The quadratic formula is derived by completing the square for the equation $ax^2 + bx + c$. Describe what is happening in each step.

Step 1: $a\left(x^2 + \dfrac{b}{a}x\right) + c = 0$

Step 2: $a\left(x^2 + \dfrac{b}{a}x + \dfrac{b^2}{4a^2} - \dfrac{b^2}{4a^2}\right) + c = 0$

Step 3: $a\left(x^2 + \dfrac{b}{a}x + \dfrac{b^2}{4a^2}\right) - \dfrac{b^2}{4a^2} + c = 0$

Step 4: $a\left(x + \dfrac{b}{2a}\right)^2 - \dfrac{b^2}{4a^2} + c = 0$

Step 5: $a\left(x + \dfrac{b}{2a}\right)^2 = \dfrac{b^2}{4a^2} - c$

Step 6: $\left(x + \dfrac{b}{2a}\right)^2 = \dfrac{b^2}{4a^2} - \dfrac{c}{a}$

Step 7: $\left(x + \dfrac{b}{2a}\right)^2 = \dfrac{b^2}{4a^2} - \dfrac{4ac}{4a^2}$

Step 8: $\left(x + \dfrac{b}{2a}\right)^2 = \dfrac{b^2 - 4ac}{4a^2}$

Step 9: $x + \dfrac{b}{2a} = \pm\dfrac{\sqrt{b^2 - 4ac}}{2a}$

Step 10: $x = \dfrac{-b}{2a} \pm \dfrac{\sqrt{b^2 - 4ac}}{2a}$

Step 11: $x = \dfrac{-b \pm \sqrt{b^2 - 4ac}}{2a}$

2. **(a)** State the discriminant for the quadratic formula if the equation is $ax^2 + bx + c = 0$.

   (b) When you use the quadratic formula to solve an equation, why does it help to find the discriminant first?

   (c) How does the discriminant of the equation reflect upon the graph of the corresponding function and its x-intercepts?

3. Verify to determine if the expressions are roots of the given quadratic equations.

Root   Equation

(a) $\dfrac{3+\sqrt{11}}{2}$ $2x^2 - 6x - 1 = 0$

(b) $\dfrac{2+\sqrt{11}}{2}$ $2x^2 - 5x - 3 = 0$

(c) $\dfrac{-5-\sqrt{7}}{4}$ $x^2 = 2x - 1$

(d) $\dfrac{\sqrt{3}}{4}$ $2x^2 + 8x + 2 = 0$

4. Find the zeros.

(a) $f(x) = x^2 + 9x + 15$ **(b)** $f(x) = 5x^2 - 8x - 3$

(c) $f(x) = 3x^2 + 6x - 24$ **(d)** $f(x) = 5x^2 + 6x + 9$

5. Find the x-intercepts.

(a) $f(x) = x^2 + 10x - 5$ **(b)** $f(x) = 3x^2 + 3x - 4$

6. Write an equation for each root.
Then write the corresponding function.

(a) $3 \pm \sqrt{5}$ **(b)** $2 \pm \sqrt{6}$

(c) $\dfrac{1 \pm \sqrt{5}}{3}$ **(d)** $\dfrac{-4 \pm \sqrt{3}}{3}$

(e) $\dfrac{3 \pm \sqrt{23}}{4}$

(f) $\dfrac{-8 \pm \sqrt{7}}{7}$

7. Why is more than one equation possible for any root?

8. When you use the quadratic formula to solve the equation $3x^2 + 5x - 3 = 0$, the result is

$\dfrac{-5 \pm \sqrt{-11}}{6}$.

    (a) What happens when you use a calculator to find $\sqrt{-11}$?

    (b) Why can't a negative number have a square root?

    (c) Why does an equation have no roots when the discriminant is negative? Then apply the quadratic formula. What happens?

$$\left[x = \dfrac{5 \pm \sqrt{-11}}{-6} \right]$$

9. Find two numbers with a difference of 1 whose product is 246.24. Write an equation and then use the quadratic formula to solve the problem.

10. The area of a square doubles when 10 in. is added to one side and 12 in. to the other. Find the dimensions of the square.

In this lesson, you learned how to use the **discriminant** in the **quadratic formula** to determine whether a quadratic equation has two roots, one root, or no roots. The discriminant is the expression $b^2 - 4ac$, which is found under the **radical** in the formula.

Recall:
- When $b^2 - 4ac > 0$, the equation has two real roots and the graph meets the x-axis in two places.
- When $b^2 - 4ac = 0$, the equation has one real root and the graph meets the x-axis in one place.
- When $b^2 - 4ac < 0$, the equation has no real roots and the graph does not meet the x-axis.

Example 1

Determine the nature of the roots for $3x^2 + 6x + 2 = 0$.

Solution

Remember that $3x^2 + 6x + 2 = 0$ is in the general form $ax^2 + bx + c = 0$.

The coefficients for the equation are $a = 3$, $b = 6$, and $c = 2$.

Substitute these values into the discriminant and evaluate.

$$b^2 - 4ac = (6)^2 - 4(3)(2)$$
$$= 36 - 24$$
$$= 12$$

The value of the discriminant, 12, is greater than 0.

Since $b^2 - 4ac > 0$, the equation $3x^2 + 6x + 2 = 0$ has two real roots.

Example 2

Find the sum and product of the roots for $3x^2 + 6x + 2 = 0$, without finding the roots.

Solution

The sum of the roots is $-\frac{b}{a}$ and the product of the roots is $\frac{c}{a}$.

Substitute $a = 3$, $b = 6$, and $c = 2$ into each expression.

Sum of the roots $= -\frac{b}{a}$ product of the roots $= \frac{c}{a}$

$$= -\frac{6}{3} \qquad\qquad = \frac{2}{3}$$
$$= -2$$

The sum of the roots is -2 and the product of the roots is $\frac{2}{3}$.

Exercises

1. Why is it useful to begin solving an equation by expressing it in the form
   $ax^2 + bx + c = 0$?

2. When you use the quadratic formula to solve an equation, why does it help to
   evaluate the discriminant first?

3. What does the discriminant of the equation tell you about the *x*-intercepts of the
   corresponding graph?

4. Why can an equation not be factored if the value of the discriminant is negative?

5. Determine the number of real roots for each graph.

   (a)

   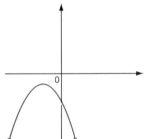

   (b)

   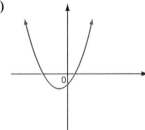

   (c)

   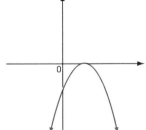

6. Determine the number of real roots for each equation.

(a) $x^2 - 4x - 6 = 0$

(b) $x^2 - 10x + 25 = 0$

(c) $x^2 + 3x + 8 = 0$

(d) $15 - 11x = 12x^2$

(e) $11x = 7x^2 + 5$

(f) $49 + 42x = 9x^2$

7. Determine the value(s) of k for each equation.

(a) $9x^2 + 12x + k = 0$, if there is one real root

(b) $7x^2 + kx + 8 = 0$, if there are no real roots

(c) $3x^2 - 5x + 4k = 0$, if there are two real roots

(d) $-3 = kx^2 + 6x$, if there are two real roots

(e) $10kx = -16x^2 - 25$, if there is one real root

(f) $5x + k = 2x^2$, if there are no real roots

8. Find the sum and product of the roots for each equation without solving for the roots.

   (a) $x^2 - 6x + 13 = 0$

   (b) $4x^2 + 6x - 7 = 0$

   (c) $-3x^2 + 14x = -8$

9. Determine the value(s) of n for each equation.

   (a) $-x^2 + nx + 40 = 0$, where one root is 6 less than the other root

   (b) $9x^2 + 2nx + 8 = 0$, where one root is double the other root

10. Two numbers have a sum of 10. Determine if the product of these two numbers can be equal to 30.

11. A rocket is fired on a practice range. The height, h (in meters), of the rocket after t seconds is found using the equation: $h = -\frac{1}{4}t^2 + 5t + 105$.

    Determine if the rocket can reach a height of 120 m.

12. Create a quadratic equation that has
    (a) two real roots   (b) one real root   (c) no real roots

    Exchange equations with a classmate. Solve and graph each equation.

10.6 Graphs of Polynomial Functions

In this lesson, you learned how to graph, describe, and analyze polynomial functions in one variable.

- For an **even-degree** polynomial function, both arms of the graph point in the same direction. If the **leading coefficient** is positive, the arms point up; if it is negative, the arms point down.
- For an **odd-degree** polynomial function, the arms of the graph point in opposite directions. If the leading coefficient is positive, the leftmost points of the graph have negative y-values and the rightmost points have positive y-values. If the leading coefficient is negative, these characteristics are reversed.
- The maximum number of changes in direction of the graph is one less than the degree of the function.
- The sum of the **multiplicities** of all zeros, real and imaginary, must equal the degree of the function.
- If the multiplicity of a real zero is odd, the function crosses the x-axis at that point.
- If the multiplicity of a real zero is even, the function touches the x-axis at that point.

Example 1

Given the function $f(x) = -x^4 + 3x^3 + x^2 - 5x + 2$, identify the following characteristics:

(a) leading coefficient

(b) degree

(c) constant term

(d) maximum number of changes in direction of the graph

(e) direction in which the arms of the graph point

Solution

(a) The leading coefficient is the coefficient of the term with the greatest degree, $-x^4$. The coefficient of the term is -1.

(b) The degree of a function is defined by the term with the greatest degree. The degree of $-x^4$ is 4.

(c) The constant term is 2. The constant term is also called the y-intercept.

(d) The maximum number of changes in direction is one less than the degree of the function: $4 - 1 = 3$

(e) Even-degree functions have graphs with arms that point in the same direction. The negative leading coefficient indicates that the arms point down.

$y = -x^4 + 3x^3 + x^2 - 5x + 2$

Example 2

Given the graph of the cubic function, identify its characteristics.

(a) degree

(b) sign of the leading coefficient

(c) constant term

(d) real zeros

(e) multiplicity of the zeros

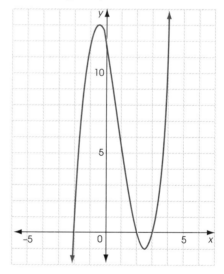

Solution

(a) All cubic functions have a degree of 3.

(b) Since the rightmost points have positive y-values, the sign of the leading coefficient must be positive.

(c) The graph appears to cross the y-axis at 12, so the constant term is 12.

(d) Real zeros occur where the graph crosses the x-axis. The zeros are at -2, 2, and 3.

(e) A cubic function has three zeros. Since the graph crosses the x-axis three times, each zero has a multiplicity of 1.

Exercises

1. Explain each term in your own words.

   (a) leading coefficient

   (b) multiplicity of zeros

   (c) zero of a polynomial function

   (d) degree of a polynomial function

   (e) quartic function

   (f) quintic function

2. What effect does the sign of the leading coefficient have on the graph of a polynomial function? Explain in your own words.

3. Describe the relationship between the degree of a polynomial function, the number of changes in direction of the function's graph, and the direction in which the arms point.

4. For each polynomial function, state the leading coefficient (a_n), the degree (n), and the constant term (a_0).

   (a) $y = x^5 - 2x^4 + 3$

   (b) $f(x) = 7x^2 + 3x^5 + 8$

   (c) $f(x) = -2x^3 + 7x^2 + x$

   (d) $y = (x-1)^4$

   (e) $f(x) = -(x-2)^2(x+1)^3$

5. Match each equation to its graph.

(a) $f(x) = x^3 - 5x$

(b) $f(x) = -x^4 + x + 4$

(c) $f(x) = -x^3 + 2x^2$

(d) $y = -x^5 + 3x^3 + 1$

(e) $y = x^5 - 3x^3 + 2$

(f) $y = x^4 - 4x^2 + x$

(i)

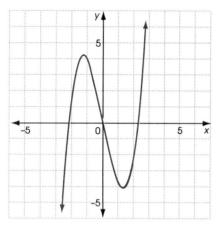

(ii)

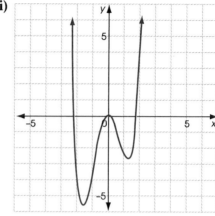

(iii)

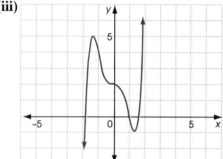

(iv)

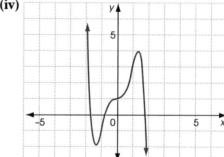

(v)

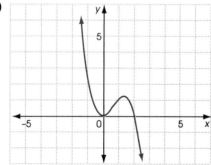

(vi)

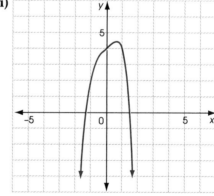

6. The following partial table of values is given for graph (iv) in Problem 5. Estimate the positive zero.

x	f(x)
9	
-1	
1	
3	
-7	

7. State the number of different zeros in graph (ii) of Problem 5 and the multiplicity of each zero.

8. Given the function $y = -(x-2)^2(x-4)$:
   (a) What is the y-intercept of the graph?

   (b) What are the zeros of the function?

   (c) What is the multiplicity of each zero?

   (d) Does the graph eventually fall or rise to the right?

   (e) Do the arms of the graph point in the same or opposite directions?

9. Write several polynomial functions varying the degree, the sign of the leading coefficient, and the y-intercept (constant term). Challenge classmates to describe the shape of each graph and the directions in which the arms point. You could use a graphing tool to help create equations and to check work.

10.7 Graphs of Rational Functions

You have learned to graph rational functions and to describe and analyze the graphs using the correct terminology.

- A rational function is a function of the form $y = \dfrac{P(x)}{Q(x)}$, where both $P(x)$ and $Q(x)$ are polynomials and $Q(x) \neq 0$.
- To find **vertical asymptotes**, determine the restrictions on $Q(x)$. Include these restrictions when determining the **domain** of the function.
- To find a **horizontal asymptote**, study the pattern of values of y as $|x|$ increases. Include this restriction when determining the **range** of the function.

Example 1

Describe the graph by identifying the domain, range, intercepts, and asymptotes.

$$y = \frac{1}{x-3}$$

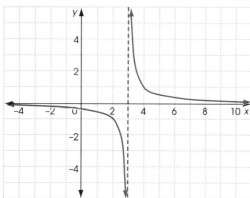

Solution

The line $x = 3$ is the vertical asymptote, so 3 must be excluded from the domain. The domain is $\{x | x \neq 3,\ x \in \mathrm{R}\}$.

The line $y = 0$ is the horizontal asymptote, so 0 must be excluded from the range. The range is $\{y | y \neq 0,\ y \in \mathrm{R}\}$.

There are no x-intercepts. To find the y-intercept, substitute 0 for x in the equation and solve for y.

$$y = \frac{1}{0-3}$$
$$= -\frac{1}{3}$$

The y-intercept is $-\dfrac{1}{3}$.

Example 2

Sketch the graph defined by $f(x) = \dfrac{1}{3x-6}$.

Solution

If the degree of $P(x)$ is less than the degree of $Q(x)$, then the horizontal asymptote is $y = 0$.

If the degree of $P(x)$ is equal to the degree of $Q(x)$, then the horizontal asymptote is $y = \dfrac{a}{b}$, where a and b are the leading coefficients of $P(x)$ and $Q(x)$, respectively.

If the degree of $P(x)$ is greater than the degree of $Q(x)$, then there is no horizontal asymptote.

Find the asymptotes.

The restrictions on x determine the vertical asymptote(s).

$$3x - 6 = 0$$
$$3x = 6$$
$$x = 2$$

$x = 2$ is the equation of the vertical asymptote. Draw a dotted vertical line at $x = 2$.

The numerator of the function is a constant, so the degree of $P(x)$ is 0, which is less than the degree of $Q(x)$. Therefore, the horizontal asymptote is the line $y = 0$.

Find the intercepts.

The graph does not cross the x-axis; therefore, there are no x-intercepts.

To find the y-intercept, set $x = 0$ and solve for y.

$$f(x) = \frac{1}{3x-6} \Rightarrow y = \frac{1}{3x-6}$$
$$y = \frac{1}{3(0)-6}$$
$$y = -\frac{1}{6}$$

Sketch the graph of the function.

Develop a table of values, plot the points, and sketch the graph.

x	y
-1	$-\dfrac{1}{9}$
0	$-\dfrac{1}{6}$
1	$-\dfrac{1}{3}$
3	$\dfrac{1}{3}$
4	$\dfrac{1}{6}$

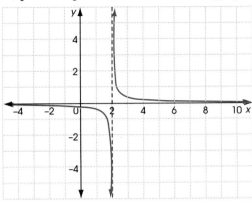

Exercises

1. Define each term.
   (a) domain
   (b) range

   (c) asymptote
   (d) point of discontinuity

2. Explain what it means to determine the restrictions on x.

3. Which of the rational functions are polynomial functions?
   (a) $y = x - 2$
   (b) $y = (x + 4)^{-1}$
   (c) $f(x) = \dfrac{x + 2}{x - 6}$

   (d) $y = \dfrac{1}{4}x^2 + 1$
   (e) $f(x) = \dfrac{x - 1}{4}$
   (f) $y = (x - 3)^{-2}$

4. Find the vertical asymptote for each function.
   (a) $f(x) = \dfrac{1}{x - 6}$
   (b) $f(x) = \dfrac{3}{2x - 1}$
   (c) $f(x) = \dfrac{-3}{x}$

(d) $f(x) = \dfrac{7}{x}$ **(e)** $f(x) = \dfrac{-3}{x^2}$ **(f)** $f(x) = \dfrac{3x+1}{x}$

5. Find the horizontal asymptote for each function.

(a) $y = \dfrac{1}{2x-1}$ **(b)** $f(x) = \dfrac{3x}{x-2}$ **(c)** $y = \dfrac{x}{x-1}$

(d) $f(x) = \dfrac{-3}{x+5}$ **(e)** $y = \dfrac{1}{3x-6}$ **(f)** $y = \dfrac{1}{x^2-4}$

6. Complete the table of values for each rational function, and sketch the graph.

(a) $f(x) = \dfrac{6}{x}$

x	y
-10	
-5	
-1	
0	
1	
5	
10	

(b) $y = \dfrac{-12}{x}$

x	y
-10	
-5	
-1	
0	
1	
5	
10	

(c) $y = \dfrac{x+3}{x-4}$

x	y
-4	
-2	
-1	
0	
2	
4	
6	

(d) $f(x) = \dfrac{x}{x+6}$

x	y
-10	
-6	
-1	
0	
1	
4	
6	

7. Use a graphing calculator to graph each rational function. From the graph, determine its domain and range and its axes of symmetry, if any.

(a) $f(x) = \dfrac{x}{x-2}$

(b) $y = \dfrac{x+2}{x}$

(c) $f(x) = \dfrac{1}{x^2 - 4}$

(d) $y = \dfrac{x+1}{x^2 - 3x - 4}$

> Use these rules to determine axes of **symmetry**. If an equivalent equation results when:
> - x and y are interchanged, then the graph is symmetric with respect to the line $y = x$.
> - x is replaced with $-x$ and y is replaced with $-y$, then the function is an odd function and is symmetric with respect to the origin.
> - x is replaced with $-x$, then the function is an even function and is symmetric with respect to the y-axis.

8. Complete the chart that describes the graph of the function $f(x) = \dfrac{x+3}{x-2}$.

vertical asymptote	
horizontal asymptote	
domain	
range	
x-intercept	
y-intercept	

9. Complete the chart that describes the graph of the function $f(x) = \dfrac{x+3}{x^2 + x - 6}$.

vertical asymptote(s)	
horizontal asymptote	
domain	
range	
x-intercept	
y-intercept	

SOLVING EQUATIONS

11.1 Absolute Value Equations

In this lesson, you learned two ways to solve absolute value equations.

1. Algebraically: Consider each case separately.
- Isolate the absolute value expression, if necessary.
- Replace the absolute value equation with two equations not containing absolute value (one which assumes that the absolute value expression is positive and one which assumes it is negative).
- Solve each equation.
- Verify each solution and reject extraneous roots.

2. Graphically: Graph $y = $ *left side* and $y = $ *right side* of the absolute value equation and find the point(s) of intersection.

Example 1
Solve for x. $|x - 6| = 2x$

Solution
$|x - 6| = 2x$

Case 1	**Case 2**	
$x - 6 = 2x$	$x - 6 = -2x$	Replace the absolute value equation with two equations not containing absolute value.
$x = -6$	$x = 2$	Solve each equation for x.

Verify each solution.

If $x = -6$:			If $x = 2$:	
L.S.	R.S.		L.S.	R.S.
$\|x - 6\|$	$2x$		$\|x - 6\|$	$2x$
$= \|6 - 6\|$	$= 2(6)$		$= \|2 - 6\|$	$= 2(2)$
$= \|0\|$	$= 12$		$= \|-4\|$	$= 4$
$= 0$	$= 12$		$= 4$	$= 4$
L.S. $\neq$ R.S.			L.S. $=$ R.S.	

Discard the extraneous root $x = -6$. The only solution is $x = 2$.

Example 2
Solve for x. $2|x + 3| + 5 = 15$

Solution

$2|x + 3| + 5 = 15$	Isolate the absolute value expression.
$2|x + 3| = 10$	
$|x + 3| = 5$	

Case 1	**Case 2**	
$x + 3 = 5$	$x + 3 = -5$	Replace the absolute value equation with two not containing absolute value.
$x = 2$	$x = -8$	Solve each equation for x.

Verify each solution.

If $x = 2$:		If $x = -8$:	
L.S.	R.S.	L.S.	R.S.
$2\|x + 3\| + 5$	15	$2\|x + 3\| + 5$	15
$= 2\|2 + 3\| + 5$		$= 2\|-8 + 3\| + 5$	
$= 2\|5\| + 5$		$= 2\|-5\| + 5$	
$= 2(5) + 5$		$= 2(5) + 5$	
$= 10 + 5$		$= 10 + 5$	
$= 15$		$= 15$	
L.S. $=$ R.S.		L.S. $=$ R.S.	

Since the left side equals the right side for both $x = 2$ and $x = -8$, both solutions verify.

Exercises

1. Explain the meaning of absolute value. Why is it necessary to consider two cases?

2. Why do you think absolute value graphs are V-shaped? Do you think all absolute value graphs look like this?

3. Simplify.
   (a) $|8|$

   (b) $|-14|$

4. Simplify.
   (a) $|6| + |-3|$
   (b) $|-4| - |-9|$
   (c) $2|-6| + 3|-2|$

5. Evaluate the following.
   (a) $|x - 3|$ for $x = -5$

   (b) $|2a + 6|$ for $a = -4$

   (c) $3|y - 5|$ for $y = 2$

   (d) $-2|3x + 7| - 6$ for $x = -3$

6. Write the definition of $|x|$ in mathematical symbols.

7. Solve and verify.
   (a) $|x| = 3$

   (b) $|x + 2| = 9$

   (c) $|2x + 3| = 7$

   (d) $|x - 5| - 8 = 0$

   (e) $-2|x + 1| + 8 = 0$

   (f) $-|2x - 3| + 7 = 0$

8. Solve and verify.

    (a) $|x - 12| = 2x$

                                                 (b) $2|x + 5| = |x + 5| + 4$

    (c) $|x + 3| = -8$

                                                 (d) $|2x - 10| + |x - 5| = 9$

    (e) $|7 - x| = 9$

                                                 (f) $|4x - 8| - |6x - 12| - 2 = 0$

9. Examine the graphs to solve the equations.

    (a) $|x + 6| = |x|$

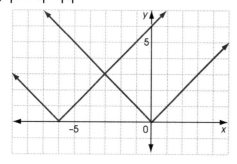

                                                 (b) $2|x - 2| = |x + 1|$

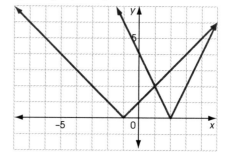

    (c) $|x + 3| = |x + 1|$

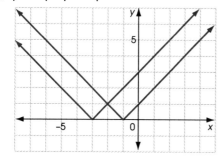

                                                 (d) $|x + 2| = |x + 1| + 2$

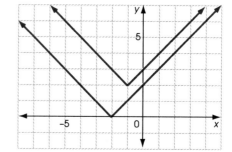

10. Carol tried to solve $|x - 8| = 3x$. Correct any errors in the solution.

$$|x - 8| = 3x$$

Case 1

$$x - 8 = 3x$$
$$2x = -8$$
$$x = -4$$

Case 2

$$x - 8 = -3x$$
$$-4x = -8$$
$$x = 2$$

Solution set = {−4, 2}

11. Use a table of values to find points on $x = |y|$, then graph the relation on a coordinate grid. How is this graph similar to the graph for $y = |x|$? Using this information, try to graph $x = |y + 1|$ without using a table of values.

11.2 Radical Equations

You have learned how to solve radical equations of various types. To solve a radical equation:

1. Isolate the radical.

2. Square both sides of the equation.

3. Simplify and solve for x.

4. Check by substitution to see whether any roots are extraneous.

For problems where you need to square more than once (with nested radicals and where there is more than one radical in the equation), remember to isolate the radical at each step.

Example 1

Solve.

$$\sqrt{x-2} - \sqrt{2x+3} = -2$$

Solution

$$\sqrt{x-2} - \sqrt{2x+3} = -2$$

❶ $$\sqrt{x-2} = \sqrt{2x+3} - 2$$

❷ $$x-2 = 2x+3 - 4\sqrt{2x+3} + 4$$

❸ $$x-2 = 2x+7 - 4\sqrt{2x+3}$$

❹ $$4\sqrt{2x+3} = x+9$$

❺ $$16(2x+3) = x^2 + 18x + 81$$

❻ $$32x+48 = x^2 + 18x + 81$$

❼ $$0 = x^2 - 14x + 33$$

❽ $$0 = (x-3)(x-11)$$

$$x = 3 \text{ and } x = 11$$

❶ Isolate a radical.
❷ Square both sides of the equation.
❸ Collect the like terms on the right side.
❹ Isolate the radical on the left side.
❺ Square both sides of the equation.
❻ Simplify.
❼ Collect like terms.
❽ Factor.

Check $x = 3$.

L.S.	R.S.
$\sqrt{x-2} - \sqrt{2x+3}$	-2
$= \sqrt{11-2} - \sqrt{2(11)+3}$	
$= \sqrt{9} - \sqrt{25}$	
$= 3 - 5$	
$= -2$	

L.S. = R.S.
Therefore, 3 is a root.

Check $x = 11$.

L.S.	R.S.
$\sqrt{x-2} - \sqrt{2x+3}$	-2
$= \sqrt{3-2} - \sqrt{2(3)+3}$	
$= \sqrt{1} - \sqrt{9}$	
$= 1 - 3$	
$= -2$	

L.S. = R.S.
Therefore, 11 is a root.

Example 2

Solve.

$$\sqrt{x+4} + 2 = x$$

Solution

$$\sqrt{x+4} + 2 = x$$

❶ $$\sqrt{x+4} = x - 2$$

❷ $$x+4 = x^2 - 4x + 4$$

❸ $$0 = x^2 - 5x$$

❹ $$0 = x(x-5)$$

$$x = 0 \text{ and } x = 5$$

❶ Isolate the radical.
❷ Square both sides of the equation.
❸ Collect like terms.
❹ Factor.

Check $x = 0$.

L.S.	R.S.
$\sqrt{0+4} + 2$	0
$= \sqrt{4} + 2$	
$= 2 + 2$	
$= 4$	

L.S. ≠ R.S.
Therefore, 0 is an extraneous root.

Check $x = 5$.

L.S.	R.S.
$\sqrt{5+4} + 2$	5
$= \sqrt{9} + 2$	
$= 3 + 2$	
$= 5$	

L.S. = R.S.
Therefore, 5 is a root.

Exercises

1. Explain what is meant by the expression "isolate the radical."

2. Explain what is meant by the expression "extraneous root."

3. Explain why you have to check the roots you have found for a radical equation.

4. By inspection, find the value of the variable that will make each equation true.

   (a) $\sqrt{a} = 7$

   (b) $\dfrac{10}{\sqrt{x}} = 5$

   (c) $-2\sqrt{x} = -16$

   (d) $3\sqrt{x} = -9$

5. Write the first step in solving each equation.

   (a) $\sqrt{4-x} - \sqrt{x+6} = 2$

   (b) $\sqrt{\sqrt{x^2 - 5}} = 4$

   (c) $\sqrt{x+2} - 3 = 9$

   (d) $5x = 1 + \sqrt{3 - 2x}$

   (e) $\dfrac{\sqrt{x+2}}{3} = 4x$

6. Solve.

   (a) $\dfrac{15}{\sqrt{x}} = 5$

   (b) $\dfrac{8}{\sqrt{x+2}} = 4$

7. Solve.

   (a) $\sqrt{4-x} - \sqrt{x+6} = 2$

   (b) $\sqrt{\sqrt{x^2 - 33}} = 4$

   (c) $\sqrt{x+2} - 3 = 9$

   (d) $5x = 1 + \sqrt{7 - 5x}$

(e) $\dfrac{\sqrt{x+2}}{3} = \dfrac{x}{3}$

8. The area of a rhombus can be found with the formula $A = \dfrac{D_1 \cdot D_2}{2}$, where D_1 and D_2 are the lengths of the diagonals. If the area of a rhombus is 9 in.^{2} and the lengths of the diagonals are given as $(\sqrt{x}+2)$ in. and $(\sqrt{x}-1)$ in., what are the actual lengths of the diagonals?

9. Police can estimate how fast a car was traveling before an accident by measuring its skid marks and using the formula $s = \sqrt{30df}$, where s is the speed of the car in miles per hour, d is the length of the skid marks in feet, and f is the coefficient of friction, which depends on road and weather conditions. The coefficient for dry pavement is approximately 1.0, while the coefficient for wet pavement is about 0.5.
(a) Why is the coefficient of friction lower for wet pavement?

(b) Complete the chart below and tell which accidents occurred under wet conditions and which occurred under dry conditions. Explain.

Accident number	Speed of vehicle (mph)	Length of skidmarks (ft)	Coefficient of friction
1	55		0.56
2		144	0.87
3	50	172	
4		107	0.34
5	64		0.24
6		215	0.59
7	40	192	
8	32	40	
9	75		0.23

(c) A car left 150 ft skid marks. What was its speed if the pavement was dry? wet?

(d) One set of skid marks is twice as long as a second set. Do you think that the second car was traveling twice as fast as the first car? If not, was it going less than twice as fast or more than twice as fast? Explain.

(e) A car is traveling at 65 mph on concrete. How long will its skid marks be if the concrete is dry ($f = 0.82$)? wet ($f = 0.42$)?

(f) You have been invited to give a lecture called "Skidding Distance and Investigating Highway Accidents" to a group of cadets who have been selected to join the highway patrol. Prepare an illustrated two- or three-page report including all pertinent information and diagrams. A poster may be helpful.

10. Create and solve a radical equation where one of the roots is an extraneous root.

11.3 Rational Equations

In this lesson, you learned how to solve rational equations.
The steps for solving a rational equation are:

1. Determine any non-permissible values.

2. Multiply both sides of the equation by the LCD.

3. Divide out common factors.

4. Factor the simplified equation.

5. Solve the simplified equation.

6. Write the solution set.

Example

Solve.

$$\frac{2x-9}{x-7}+\frac{x}{2}=\frac{5}{x-7}$$

Solution

$$\frac{2x-9}{x-7}+\frac{x}{2}=\frac{5}{x-7} \qquad x \neq 7$$

Determine any non-permissible values. $x-7$ is the only denominator that involves the unknown.
$x-7=0$ when $x=7$, so the non-permissible value for this equation is $x=7$.

$$LCD = 2(x-7)$$

The factors which make up the two denominators are $x-7$ and 2. There are no repeated factors, so the LCD is their product, $2(x-7)$.

$$2(\cancel{x-7})\frac{2x-9}{\cancel{x-7}}+\cancel{2}(x-7)\frac{x}{\cancel{2}}=2(\cancel{x-7})\frac{5}{\cancel{x-7}}$$

Multiply both sides of the equation by the LCD.

$$2(2x-9)+(x-7)(x)=2(5)$$

Divide out common factors.

$$4x-18+x^2-7x=10$$

Multiply to remove parentheses

$$x^2-3x-28=0$$

Simplify and collect like terms.

$$(x-7)(x+4)=0$$

Factor.

$$x=7 \text{ or } x=-4$$

Solve.

Since $x=7$ is a non-permissible value, the solution set includes only $x=-4$.

Write the solution set.

Solution set = {–4}

Exercises

1. Explain how to find the lowest common denominator for Problem 3(b).
   Why is it easiest to use the *lowest* common denominator?
   What will happen if you don't use the lowest common denominator?

2. Why is it important to find non-permissible values of rational equations?
   On a graph of a rational function, what happens at the non-permissible values?

3. Determine the non-permissible values for each equation.

   (a) $\dfrac{1}{x-1} + \dfrac{x+1}{4-x} = \dfrac{2}{x}$

   (b) $\dfrac{-2}{x^2-4} + \dfrac{1}{x} = \dfrac{2}{x-2}$

   (c) $\dfrac{2x-3}{3x^2-x-4} = 1-x$

   (d) $\dfrac{2x}{x^2-9} + 2 = \dfrac{x-3}{x^2+7x+12}$

   (e) $-3 + \dfrac{4-x}{2x^2+x-3} = \dfrac{x-1}{6x^2+11x+3}$

4. Determine the LCD for each equation in Problem 3.

5. Solve each equation and state its solution set.

(a) $3 - \dfrac{2}{x} = \dfrac{2}{x-1}$

(b) $\dfrac{x}{x-4} - \dfrac{1}{x-3} = \dfrac{x}{x^2 - 7x + 12}$

(c) $\dfrac{x}{x+1} - \dfrac{1}{x+2} = \dfrac{1}{x^2 + 3x + 2}$

(d) $2x + \dfrac{4x - 11}{x+1} = 1$

(e) $\dfrac{x}{x+1} + \dfrac{x-11}{x+6} = \dfrac{7x-5}{x^2 + 7x + 6}$

6. The Waugh family is driving from their home in Wyoming to visit relatives in California, a distance of 780 mi. The round trip takes 25 hours. Their speed on the way to California is 5 mph faster than on their way home. What is their speed on the way home?

7. Each equation was solved incorrectly. Find the error in each solution, and complete the solution to find the correct answers.

(a)
$$\dfrac{3x^2}{3x+1} - 2 = \dfrac{2x-1}{3x+1}$$
$$(3x+1)\dfrac{3x^2}{3x+1} - 2 = (3x+1)\dfrac{2x-1}{3x+1}$$
$$3x^2 - 2 = 2x - 1$$
$$3x^2 - 2x - 1 = 0$$
$$(3x+1)(x-1) = 0$$

$$x = -\dfrac{1}{3} \text{ or } x = 1$$

Solution set = {1}

(b)

$$\frac{1}{1-x} = 1 - \frac{2x}{x-1}$$

$$(x-1)\frac{1}{1-x} = 1(x-1) - \frac{2x}{x-1}(x-1)$$

$$1 = x-1-2x$$

$$2 = -x$$

$$x = -2$$

Solution set = {−2}

(c)

$$\frac{8}{x+3} - \frac{6}{x-1} = -3$$

$$(x+3)(x-1)\frac{8}{x+3} - (x+3)(x-1)\frac{6}{x-1} = -3(x+3)(x-1)$$

$$8x-8-6x+18 = -3x^2 - 6x - 9$$

$$3x^2 + 8x + 19 = 0$$

$$x = \frac{-3 \pm \sqrt{3^2 - 4(3)(19)}}{2(3)} = \frac{-3 \pm \sqrt{-219}}{6}$$

Solution set = Ø

8. To the right of an equation, you see the following: $x > -2$, $x \neq 1$.
   State the non-permissible values for this equation.

9. Find a formula in your physics textbook that involves rational expressions.
   Create a problem using this formula that requires solving a rational equation.

11.4 Rational Inequalities 1

In this lesson, you learned how to solve a quadratic, polynomial, or rational inequality in one variable:

- Find the critical values of the inequality.
- Test numbers in the regions created by the critical values.
- The solution is the region(s) resulting in a true inequality.
- Do not include the non-permissible values in the solution.

You also learned how to graph a quadratic inequality in two variables:

- Graph the corresponding equation. If the inequality is $\leq$ or $\geq$, use a solid line; if the inequality is $<$ or $>$ use a broken line. Test points above and below the parabola. Shade the region containing the point that results in a true inequality.

Example 1

Solve $x^2 + 6x - 16 \geq 0$.

Solution

Step 1

Factor.

$(x + 8)(x - 2) \geq 0$

Step 2

Equate each factor to 0 to find the critical value(s).

$x + 8 = 0$ or $x - 2 = 0$

$x = -8$ $x = 2$

Critical values $= -8, 2$

Step 3

Place the critical values on a number line and evaluate the signs of the regions created.

If $x = -9$,
$(x + 8)(x - 2)$
$= (-)(-)$
$= +$

If $x = 0$
$(x + 8)(x - 2)$
$= (+)(-)$
$= -$

If $x = 3$
$(x + 8)(x - 2)$
$= (+)(+)$
$= +$

Step 4

Since $(x + 8)(x - 2) \geq 0$, you need to identify the regions where the product of $(x + 8)$ and $(x - 2)$ is positive or 0. The solution is $\{x \mid x \leq -8 \text{ or } x \geq 2\}$ or $(-\infty, -8] \cup [2, \infty)$.

Example 2

Graph $y < x^2 + 6x - 16$.

Solution

Step 1

Find the vertex by completing the square.

$y < (x^2 + 6x + 9) - 16 - 9$

$y < (x + 3)^2 - 25$

vertex $= (-3, -25)$

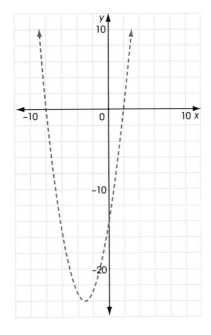

Step 2

Sketch the parabola.

Use the equation to locate a few points that are on the graph of $y < (x + 3)^2 - 25$, then sketch the graph. Since the inequality is "less than," use a broken line.

Step 3

Shade the correct region.

Since the inequality is "less than," shade the region below the parabola.

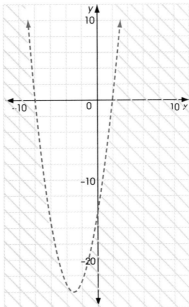

Exercises

1. Fill in the blanks to make each statement true.

   (a) Any inequality of the form $ax^2 + bx + c > 0$ $(a \neq 0)$ is called a _____ inequality.

   (b) The inequality $y < x^2 - 2x + 3$ is a non-linear inequality in _____ variables.

   (c) In interval notation, the _____ $(3, 5)$ represents the real numbers between 3 and 5.

2. Explain why $(x - 4)(x + 5)$ will be positive only when the signs of $x - 4$ and $x + 5$ are the same.

3. Graph on a separate sheet of paper.
   (a) $y > (x+4)^2 + 4$

   (b) $y \le -x^2 + 4x + 1$

   (c) $y > -3x^2 + 36x - 110$

   (d) $y \ge \frac{1}{2}x^2 + x - \frac{11}{2}$

   (e) $y \le 4x^2 - 24x + 43$

4. Convert to interval notation.
   (a) $\{x \mid -2 < x \le 5\}$

   (b) $\{x \mid x \le -6 \text{ or } x > 7\}$

   (c) $\{x \mid -2 \le x < 3 \text{ or } x = 11 \text{ or } x > 15\}$

5. Solve.
   (a) $(x-5)(x+3) > 0$

   (b) $-(x-7)(x-4) \le 0$

   (c) $x^2 + 10x + 24 \le 0$

   (d) $x^2 - 3x - 88 > 0$

   (e) $2x^2 - 19x - 10 < 0$

   (f) $18x - 120 \ge -3x^2$

   (g) $20 - 11x \ge 3x^2$

6. Solve.
   (a) $(x+2)(x-2)(x+5) > 0$

   (b) $(x-5)(x+3)(x-7) \le 0$

   (c) $(x-8)(x-2)(x-11) \ge 0$

   (d) $-(x+3)^2(x+1)(x-6)^3 < 0$

   (e) $(x^2 - 36)(x-8)^2 \le 0$

   (f) $5x^2(x^2 + 2x - 35) < 0$

   (g) $(x+2)(x-8)(x+7)(x-10) \ge 0$

   (h) $(x-4)^2(x+11)^3(x-6)^5(x-9)^4 > 0$

7. Solve. Remember to exclude any non-permissible values from the solution.

(a) $\dfrac{x-3}{x} > 0$

(b) $\dfrac{x+7}{x+1} \le 0$

(c) $\dfrac{(x+5)(x-2)}{x+8} < 0$

(d) $\dfrac{(x-1)}{(x+3)(x+12)} \ge 0$

(e) $\dfrac{x^2-6x+8}{x^2+5x-24} \le 0$

(f) $\dfrac{(x+6)^2(x-2)}{4x^3(x-10)} \le 0$

(g) $\dfrac{3}{x-2} < \dfrac{4}{x}$

(h) $\dfrac{x}{x+9} \ge \dfrac{1}{x+1}$

8. The length of a desktop is to be made 30 in. longer than the width. What are possible widths of this desktop that would result in an area of at least 1,800 in.^{2}?

9. A jeweler makes silver bracelets. He has determined that if he sells the bracelets for 60 − x dollars, where x is the number of bracelets produced each week, then he is able to sell all of the bracelets made. How many bracelets must he make so that his revenue is at least $500 per week?

10. Emilio incorrectly solved this inequality. Find his error and write a correct solution.

$$\dfrac{x^2-2x-8}{x^2+5x-6} \le 0$$
$$\dfrac{(x-4)(x+2)}{(x+6)(x-1)} \le 0$$

Critical values = −4, −1, 2, 6

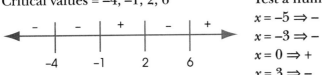

Test a number in each region.
$x = -5 \Rightarrow -$
$x = -3 \Rightarrow -$
$x = 0 \Rightarrow +$
$x = 3 \Rightarrow -$
$x = 7 \Rightarrow +$
The solution is $(-\infty, -1] \cup [2, 6]$.

Rational Inequalities 2

In this lesson, you learned how to solve absolute value and radical inequalities.

- To solve an absolute value inequality, first isolate the absolute value expression and then write two separate inequalities: one which assumes the expression is positive and one which assumes it is negative. If the original inequality is < or ≤, then the two new inequalities are separated by "and." If the original inequality is > or ≥, then the two new inequalities are separated by "or."

- To solve a radical inequality, find the restrictions on x and then solve the corresponding radical equation to find the critical values. Graph these values on a number line and label each region ∅, −, or +. Compare the inequality to 0 to identify the regions that satisfy the inequality.

Example 1

Solve $|x + 7| - 4 > 5$.

Solution

Step 1

Isolate the absolute value expression.

$|x + 7| - 4 > 5$
$\quad |x + 7| > 9$

Step 2

Write two separate inequalities.

$x + 7 > 9 \ \text{ or } \ x + 7 < -9$

Step 3

Solve each inequality.

$x + 7 > 9 \ \text{ or } \ x + 7 < -9$

$\quad x > 2 \text{ or } \ x < -16$

Step 4

Graph these inequalities on a number line.

Step 5

Write the solution.

Therefore, the solution is $(-\infty, -16) \cup (2, \infty)$.

Example 2

Solve $5 - \sqrt{x+2} \geq 3$.

Solution

Step 1
Determine the restriction of the square root.

$$x + 2 < 0$$
$$x < -2$$

Therefore, values less than –2 produce a negative radical.

Step 2
Isolate the radical expression.

$$5 - \sqrt{x+x} \geq 3$$
$$-\sqrt{x+2} \geq -2$$
$$\sqrt{x+2} \leq 2$$

Step 3
Solve the corresponding radical equation to find the critical value.

$$\sqrt{x+2} = 2$$
$$(\sqrt{x+2})^2 = 2^2$$
$$x + 2 = 4$$
$$x = 2$$

Critical value is 2.

Step 4
Relate the inequality to 0.

$$\sqrt{x+2} - 2 \leq 0$$

Step 5
Use the above inequality to evaluate the sign of the radical in each region on a number line.

Since the inequality is ≤ 0, the region that is negative contains the values that satisfy the inequality. Therefore, the solution is $[-2, 2]$.

Exercises

1. Explain the terms *and, or, conjunction,* and *disjunction,* as they apply to absolute value inequalities.

2. Define what a critical value is and how it is used in solving radical inequalities.

3. Discuss the similarities and differences between solving an absolute value equation and solving an absolute value inequality.

4. Discuss the similarities and differences between solving a radical equation and solving a radical inequality.

5. Solve.

(a) $|x| < 3$

(b) $|x - 6| \geq 5$

(c) $2|8 + x| \leq 12$

(d) $7 + |3 - x| > 18$

(e) $11 - 3|x + 5| < 38$

(f) $-5|2x| - 9 \leq -24$

(g) $14 + 3|x + 11| < 2$

(h) $4|7 - 3x| + 6 \leq 22$

6. Solve.

(a) $\sqrt{x} < 5$

(b) $\sqrt{x} + 7 \geq 11$

(c) $\sqrt{x - 4} > 2$

(d) $8 + \sqrt{x + 7} \leq 2$

(e) $4\sqrt{5 - x} > 40$

(f) $7 - \sqrt{2x + 6} \geq 1$

(g) $-4\sqrt{9 - 3x} - 11 < -3$

(h) $23 + 5\sqrt{2x - 7} \geq 58$

7. Match the absolute value inequality with its graph.

(a) $|x+2| > 1$ (i)

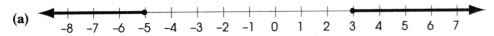

(b) $|x+2| < 1$ (ii)

(c) $|x+2| = 1$ (iii)

8. Solve by graphing the corresponding factors.

(a) $\sqrt{5-x} \le x+1$

(b) $\sqrt{x+3} > -\frac{1}{2}(x+1)^2 + 4$

(c) $2\sqrt{x-5}+1 \ge |x+1| - 5$

(d) $-\sqrt{6-x}-3 < -x^3 + 3$

9. Write an absolute value inequality for each graph.

(a)
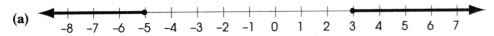

(b)
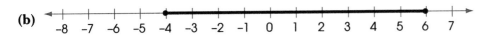

(c)
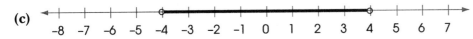

(d)
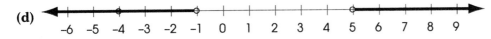

10. How does the restriction of the radical inequality $\sqrt{x-6} > 7$ aid in determining the graph of the corresponding function $y = \sqrt{x-6}$?

11. Under what conditions will the solution to an absolute value inequality be the set of all real numbers? the null set?

12. Under what conditions will a radical inequality have no solution? Why can a radical inequality never have a solution that is the set of all real numbers?

13. A car CD player has an operating temperature of $|t-40°| < 80°$, where t is a temperature in degrees Fahrenheit. Solve the inequality and express this range of temperatures as an interval.

11.6 Complex Fractions

> You have seen that there are two ways to simplify complex fractions:
> * Rewrite the fraction as the division of two separate fractions.
> * Multiply the numerator and the denominator of the complex fraction by the LCD of the individual denominators.
>
> You have also seen that factoring may be required to find the LCD and to finish the simplification of the fraction.

Example 1

Simplify.

$$\frac{\frac{3x}{5}+1}{2-\frac{x}{5}}, \quad x \neq 10$$

Solution

Using the Division Method

$$\frac{\frac{3x}{5}+1}{2-\frac{x}{5}}=\frac{\frac{3x}{5}+\frac{5}{5}}{\frac{10}{5}-\frac{x}{5}}$$

Change 1 to $\frac{5}{5}$ and 2 to $\frac{10}{5}$ in order to write the numerator and denominator as single fractions.

$$=\frac{\frac{3x+5}{5}}{\frac{10-x}{5}}$$

Express the numerator and denominator as single fractions.

$$=\frac{3x+5}{5} \div \frac{10-x}{5}$$

Write the complex fraction as an equivalent division problem.

$$=\frac{3x+5}{5} \cdot \frac{5}{10-x}$$

To divide, multiply the dividend by the reciprocal of the divisor.

$$=\frac{5(3x+5)}{5(10-x)}$$

Multiply the fractions.

$$=\frac{(3x+5)}{(10-x)}$$

Eliminate the common factor 5 from the numerator and denominator. Since there are no other factors common to the numerator and denominator, the fraction is in simplest terms.

Using the LCD Method

$$\frac{\frac{3x}{5}+1}{2-\frac{x}{5}}=\frac{5(\frac{3x}{5}+1)}{5(2-\frac{x}{5})}$$

Multiply the numerator and denominator by the LCD of the individual denominators. Since 5 is the only denominator, the LCD is 5.

$$=\frac{(3x+5)}{(10-x)}$$

Simplify the numerator and the denominator.

Exercises

1. What is a complex fraction? Give three examples.

2. When is a complex fraction in simplest terms?

3. Simplify each complex fraction.

(a) $\dfrac{\dfrac{4}{7}+\dfrac{5}{9}}{\dfrac{2}{3}+\dfrac{5}{21}}$

(b) $\dfrac{2+\dfrac{5}{11}}{3-\dfrac{3}{22}}$

4. Simplify each complex fraction.

(a) $\dfrac{\dfrac{5}{x}+5}{\dfrac{3}{2x}+2}$

(b) $\dfrac{\dfrac{9}{c-d}}{\dfrac{4}{c-d}}$

(c) $\dfrac{2-\dfrac{1}{x+y}}{3-\dfrac{2}{x+y}}$

(d) $\dfrac{\dfrac{x^2-y^2}{xy}}{\dfrac{x-y}{y}}$

(e) $\dfrac{\dfrac{1}{xy}-\dfrac{1}{y^2}}{\dfrac{1}{x^2 y}-\dfrac{1}{xy^2}}$

(f) $\dfrac{\dfrac{1}{y+2}-3}{2+\dfrac{2}{y+2}}$

(g) $\dfrac{\dfrac{1}{x-4}+5}{\dfrac{2}{x-4}+3}$

(h) $\dfrac{\dfrac{2}{n-3}-\dfrac{3}{n-2}}{\dfrac{2}{n-2}-\dfrac{3}{n-3}}$

(i) $\dfrac{\dfrac{1}{x+5}+\dfrac{1}{x-3}}{\dfrac{2x^2-3x-5}{x^2-2x-15}}$

(j) $\dfrac{x+\dfrac{xy}{y-x}}{\dfrac{y^2}{x^2-y^2}+1}$

(k) $\dfrac{\dfrac{t+6}{t}-\dfrac{1}{t+2}}{\dfrac{t^2+4t+3}{t^2+t}}$

(l) $\dfrac{\dfrac{5x}{x^2-6x+8}}{\dfrac{2}{x-4}+\dfrac{3}{x-2}}$

$$\textbf{(m)} \quad \dfrac{\dfrac{1}{y+3}+\dfrac{4y}{(y+3)^3}}{\dfrac{5}{(y+3)^2}+\dfrac{4}{y+3}}$$

$$\textbf{(n)} \quad \dfrac{\dfrac{3x-3}{x+1}}{\dfrac{x^2-1}{x^2+2x+1}}$$

5. Write Step 3 of this solution. Make sure you have completely simplified the complex fraction.

Problem

$$\frac{3+2y^{-1}-y^{-2}}{2-2y^{-2}},\ y\neq 0,\ \pm 1$$

Solution

Step 1: Rewrite the complex fraction with positive exponents.

$$\frac{3+2y^{-1}-y^{-2}}{2-2y^{-2}}=\frac{3+\dfrac{2}{y}-\dfrac{1}{y^2}}{2-\dfrac{2}{y^2}}$$

Step 2: The individual denominators are y and y^2. The LCD of these denominators is y^2. Multiply the numerator and denominator of the complex fraction by y^2.

$$=\frac{y^2\left(3+\dfrac{2}{y}-\dfrac{1}{y^2}\right)}{y^2\left(2-\dfrac{2}{y^2}\right)}$$

$$=\frac{3y^2+2y-1}{2y^2-2}$$

Step 3:

6. Simplify each complex fraction.

(a) $\dfrac{y^{-2}+1}{y^{-2}-1}$

(b) $\dfrac{2x^{-1}+4x^{-2}}{2x^{-2}+x^{-1}}$

(c) $\dfrac{x^{-2}-3x^{-3}}{3x^{-2}-9x^{-3}}$

(d) $\dfrac{1-2a^{-1}-8a^{-2}}{1-7a^{-1}+12a^{-2}}$

(e) $\dfrac{-4+10x^{-1}+24x^{-2}}{10-25x^{-1}-60x^{-2}}$

7. Create three examples of complex fractions.
   (a) Write a step-by-step solution to show how to simplify each one.

   (b) Do you think all complex fractions can be simplified? Explain why or why not.

12 EXPONENTIAL AND LOGARITHMIC FUNCTIONS

12.1 Exponential Equations

In this lesson, you learned these steps for solving **exponential equations** that have bases that are powers of one another.

- Decide on a common base to use.
- Write each side of the equation as a power with that common base.
- Once the bases are equal, the exponents are equal. Write an equation setting the exponents equal.
- Solve the resulting equation.

Example 1

Solve.

$5^{(x+6)} = 25$

Solution

$5^{(x+6)} = 25$	The left side is an exponential expression with a base of 5.
$5^{(x+6)} = 5^2$	The right side can also be written as a power with a base of 5.
$x + 6 = 2$	The bases are equal; therefore, the exponents are equal.
$x = -4$	Subtract 6 from both sides.

The solution is $x = -4$.

Example 2

Solve.

$$3^{x^2+2x} = \frac{1}{3}$$

Solution

$3^{x^2+2x} = 3^{-1}$	Write each side of the equation as an exponential expression with a base of 3.
$x^2 + 2x = -1$	The bases are equal; therefore, the exponents are equal.
$x^2 + 2x + 1 = 0$	Add 1 to both sides.
$(x+1)(x+1) = 0$	Factor the trinomial.
$x + 1 = 0$ or $x + 1 = 0$	Set each factor equal to 0.
$x = -1 \qquad x = -1$	

The solution is $x = -1$.

Example 3

Solve.

$4^x(2^{x+1}) = 16^x$

Solution

$(2^2)^x(2^{x+1}) = (2^4)^x$	Write each side of the equation as a power with a base of 2.
$2^{2x}(2^{x+1}) = 2^{4x}$	When finding a power of a power, multiply the exponents.
$2^{(2x+x+1)} = 2^{4x}$	When multiplying powers with the same base, add the exponents.
$2^{(3x+1)} = 2^{4x}$	
$3x + 1 = 4x$	The bases are equal; therefore, the exponents are equal.
$1 = 4x - 3x$	Subtract $3x$ from both sides.
$1 = x$	

The solution is $x = 1$.

Exercises

1. Explain how to solve the equation.

   $2^{x+1} = 32$

2. Which two exponential equations can be solved using the method described in this lesson? Solve the two equations.

   (a) $12^a = 24$ (b) $7^{x-3} = \dfrac{1}{49}$

   (c) $2^a = 16$ (d) $3^x = 5$

3. Solve.
   (a) $12^x = 12^6$

   (b) $5^{3x} = 5$

   (c) $10^4 = 10^{2a-6}$

   (d) $7^{x^2} = 7^{16}$

4. Check to see whether $x = -2$ is a solution of $5^{2x+3} = \dfrac{1}{5}$.

5. Solve.

    (a) $4^{2a-1} = 16$

    (b) $5^{x-4} = \dfrac{1}{125}$

    (c) $100 = 10^{2a-6}$

    (d) $3^{2x} = 81$

6. Solve.

    (a) $3^{a^2-2a} = \dfrac{1}{3}$

    (b) $3^{-3x+1} = 243$

    (c) $2^{a^2-4} = 8^a$

    (d) $25^{h-2} = 125$

    (e) $7^{x^2+3x} = \dfrac{1}{49}$

    (f) $36^{5x} = 6^x(6^{x+2})$

    (g) $32^a = 16^{3-a}$

    (h) $100^x = 0.001^{3-x}$

7. Find the two roots of the equation $4^{x^2} = 8^6$.

8. Mark solved the equation $\dfrac{4^{3x}}{4^{x+1}} = 1$, then verified the solution.

Find and correct his error.

Solution	Verification
$\dfrac{4^{3x}}{4^{x+1}} = 1$	$\dfrac{4^{3(1)}}{4^{1+1}} = 1$
$1^{2x-1} = 1^1$	$\dfrac{4^3}{4^2} = 1$
$2x - 1 = 1$	$\dfrac{64}{16} = 1$
$2x = 2$	$4 \neq 1$
$x = 1$	

9. Solve the equation $16^x = 64$ using a base of 2, and then using a base of 4.

10. Solve.

(a) $\left(\dfrac{1}{2}\right)^6 = \left(\dfrac{1}{4}\right)^{x+7}$

(b) $\left(\dfrac{1}{9}\right)^{2a} = \left(\dfrac{1}{27}\right)^{1-a}$

11. Give an example of an exponential equation that cannot be solved using the method described in this lesson.

12. Write an exponential equation that can be solved using the method described in this lesson. Make sure that different bases can be used to solve it, and write a solution. Exchange equations and solutions with a classmate and have the classmate solve the equation using a different base.

Graphing Exponential Functions

In this lesson, you learned to graph and analyze an exponential function, both with and without technology.

- An exponential function is a function of the form

    $f(x) = b^x$ or $y = b^x$

    where $b > 0$, $b \neq 1$, and x is a real number.
- A translation of the exponential function $y = a^x$ has the form

    $y = a^{(x + k)} + p$

    The constant k represents a horizontal translation by k units (left if k is positive, right if k is negative), and the constant p represents a vertical translation by p units (up if p is positive, down if p is negative).
- All exponential functions have a horizontal asymptote and no vertical asymptote.
- The graphs of $y = b^x$ and $y = \left(\dfrac{1}{b}\right)^x$ are reflections of each other over the y-axis.
- The graphs of $y = b^x$ and $y = -b^x$ are reflections of each other over the x-axis.

Example 1

Graph the exponential function $y = \left(\dfrac{3}{2}\right)^x$ using a table of values. State the domain, range, horizontal asymptote, y-intercept, and whether the function is increasing or decreasing.

Solution

Create a table of values for integral values between -4 and 4, then plot the points.

x	y
-4	$\dfrac{16}{81}$
-3	$\dfrac{8}{27}$
-2	$\dfrac{4}{9}$
-1	$\dfrac{2}{3}$
0	1
1	$\dfrac{3}{2}$
2	$\dfrac{9}{4}$
3	$\dfrac{27}{8}$
4	$\dfrac{81}{16}$

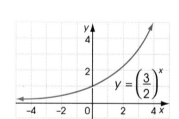

$$y = \left(\dfrac{3}{2}\right)^x$$

Since any value can be substituted for x, the domain is $(-\infty, +\infty)$. All of the points on the graph lie above the x-axis, so the range is $(0, +\infty)$. The horizontal asymptote is the x-axis, or $y = 0$. The graph crosses the y-axis at the point $(0, 1)$. As the x-values increase, the y-values also increase, so $y = \left(\dfrac{3}{2}\right)^x$ is an increasing function.

Example 2

Graph the exponential functions $y = \left(\frac{3}{2}\right)^x$ and $y = \left(\frac{2}{3}\right)^x$ on the same graph.

Determine their point of intersection. Examine the graphs. How do they relate to one another?

Solution

Add the graph of $y = \left(\frac{2}{3}\right)^x$ to the graph from Example 1.

Create a table of values for integral values between –4 and 4, then plot the points.

x	y
–4	$\frac{81}{16}$
–3	$\frac{27}{8}$
–2	$\frac{9}{4}$
–1	$\frac{3}{2}$
0	1
1	$\frac{2}{3}$
2	$\frac{4}{9}$
3	$\frac{8}{27}$
4	$\frac{16}{81}$

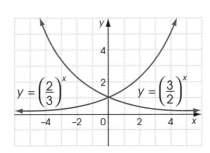

The domain, range, asymptote, and y-intercept are the same as for $y = \left(\frac{3}{2}\right)^x$. The graph of $y = \left(\frac{2}{3}\right)^x$ is a decreasing function. The graphs intersect at the y-intercept, (0, 1), and are reflections of one another over the y-axis.

Example 3

Graph the function $y = \left(\frac{1}{3}\right)^x$. Then translate the function to graph $y = \left(\frac{1}{3}\right)^{x+1} - 2$.

Check by graphing both functions on a graphing calculator.

State the intercepts of the function $y = \left(\frac{1}{3}\right)^{x+1} - 2$.

Solution

Translate the graph of $y = \left(\frac{1}{3}\right)^x$ left 1 unit and down 2

units to obtain $y = \left(\frac{1}{3}\right)^{x+1} - 2$. To determine the

intercepts, zoom in on the graph.

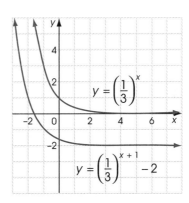

The x-intercept is (–1.63, 0), and the y-intercept is (0, –1.67).

Exercises

1. Examine the graph of $y = 3^x$.
   (a) What type of function is $y = 3^x$?
   (b) What is the domain of the function?
   (c) What is the range of the function?
   (d) What is the y-intercept of the graph?
   (e) What is the x-intercept of the graph?
   (f) What is the horizontal assmptote of the graph?
   (g) Is this an increasing or a decreasing function?
   (h) The graph passes through the point $(1, y)$. What is y?

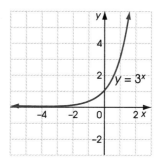

2. Graph each exponential function. State the range, horizontal asymptote, y-intercept, and whether the function is increasing or decreasing.
   (a) $f(x) = 5^x$

   (b) $g(x) = \left(\dfrac{1}{2}\right)^x$

   (c) $h(x) = 3^x - 2$
   (d) $k(x) = 2^x + 1$

   (e) $m(x) = \left(\dfrac{3}{4}\right)^x + 3$

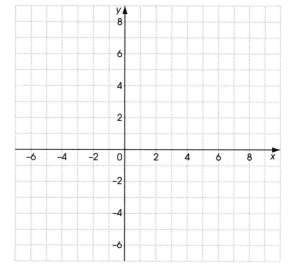

3. For each pair of exponential functions, explain how to translate the graph of the first function to obtain the graph of the second.
   (a) $y = 2^x$ \qquad $y = 2^{x+3}$

   (b) $y = 2^x$ \qquad $y = 2^{x-1} + 4$

   (c) $y = 0.5^x$ \qquad $y = 0.5^{x+1} - 2$

   (d) $y = 2^{-x}$ \qquad $y = 2^{-x-4} - 5$

4. Graph each function. State whether the function is increasing or decreasing.
   (a) $y = 3(2^{\frac{x}{4}})$

   (b) $y = -2(4^{3x})$

   (c) $y = -3\left(\dfrac{2}{5}\right)^{\frac{x}{2}}$

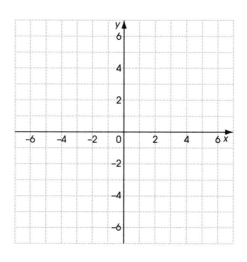

5. Reflect the graph of the exponential function $y = \left(\dfrac{2}{3}\right)^x$ over the y-axis and write two different equations for the new graph.

6. Reflect the graph of the exponential function $y = \left(\dfrac{2}{3}\right)^x$ over the x-axis and write an equation for the new graph.

7. State the y-intercept and the horizontal asymptote for each function.

   (a) $y = 5(3^x)$

   (b) $y = \dfrac{4}{3}(4^x)$

   (c) $y = -3(2^x)$

8. State the y-intercept and the horizontal asymptote for each function.

   (a) $y = 7b^x$

   (b) $y = -3b^x$

   (c) $y = b^x + 4$

   (d) $y = b^x - 2$

   (e) $y = 4\left(\dfrac{1}{b}\right)^x + 1$

9. Graph the exponential function $y = 5^x$ on your graphing calculator. Use the graph to solve each equation.

   (a) $250 = 5^x$ (b) $y = 125^{3.2}$ (c) $300 = 25^x$

Exponential Functions

In this lesson, you learned how to solve problems using exponential functions and their graphs.

Example 1

There are many different compounding periods offered by financial institutions. To compare the rates, it is necessary to determine the **effective annual interest rate** for each compounding period. If the annual interest rate is 10%, calculate the effective interest rates if interest is compounded semiannually, quarterly, monthly, and weekly.

Solution

The formula for compound amount is $A = P\left(1+\dfrac{r}{k}\right)^{kt}$, where A is the final

amount, P is the principal amount, r is the annual interest rate, k is the number of compounding periods in one year, and t is the number of years.

To find the effective interest rate, first find the final value, A, when the principal, P, is \$1 and is invested for a time, t, of 1 year.

For a \$1.00 investment for one year, the formula is $A = \left(1+\dfrac{r}{k}\right)^{k}$.

Recall that $Amount = Principal + Interest$

so $Interest = Amount - Principal$

For a \$1.00 investment, the interest earned is equivalent to the actual interest rate, so the effective interest rate for an investment can be expressed as:

$$Effective\ interest\ rate = \left(1+\frac{r}{k}\right)^{k} - P$$
$$= \left(1+\frac{r}{k}\right)^{k} - 1$$

For interest compounded semiannually,

$$Effective\ interest\ rate = \left(1+\frac{10\%}{2}\right)^{2} - 1$$
$$= (1.05)^{2} - 1$$
$$= 0.1025$$
$$= 10.25\%$$

Calculate the rates for the other interest periods.

Compounding Period	Time per Year	Effective Interest Rate
annually	1	10.000%
semiannually	2	$\left(1+\dfrac{10\%}{2}\right)^{2} - 1 = 10.250\%$
quarterly	4	$\left(1+\dfrac{10\%}{4}\right)^{4} - 1 = 10.381\%$
monthly	12	$\left(1+\dfrac{10\%}{12}\right)^{12} - 1 = 10.471\%$
weekly	52	$\left(1+\dfrac{10\%}{52}\right)^{52} - 1 = 10.506\%$

Example 2

The current population of a town is 45,000 and is increasing according to the formula:

$$P = 45{,}000(1 + r)^t$$

If the rate of increase is 7.5% per year, how long will it take for the population to double?

Solution

Double the current town's population of 45,000 is 90,000. Graph the function $P = 45{,}000(1 + 0.075)^t$ on your graphing calculator. Zoom in until you find a point having a *y*-value close to 90,000. The corresponding *x*-value is the number of years it will take for the town's population to double, approximately 9.6 years.

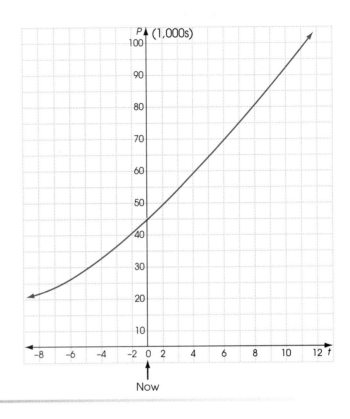

Exercises

1. An initial deposit of $7,500 earns 4% interest, compounded quarterly. How much will be in the account after five years?

2. Jamal is saving for a trip to Europe in five years. How much should he invest today, at 8% interest, compounded semiannually, if he wants to have $6,000 for his trip?

3. Which savings bond provides the best investment?
   Series A: 6.50% interest, compounded monthly
   Series B: 6.55% interest, compounded quarterly
   Series C: 6.60% interest, compounded semiannually

4. The population of a western city has been growing at a rate of 7% per year for the last 15 years. Population growth can be approximated by the formula $A = A_0(1 + r)^t$. If the current population is 750,000, what was the population 15 years ago?

5. For the years 1990–1996, a cellular telephone company experienced "exponential growth." The exponential function $S = 5.28(1.43)^t$ approximates the number of cellular telephone subscribers in thousands, where t is the number of years since 1990 and $0 \le t \le 6$.
   (a) How many subscribers were there in 1990?

   (b) How many subscribers were there in 1996?

   (c) Do you think this growth pattern will continue for the next 50 years? Why or why not?

6. A colony of six million bacteria is growing in a culture medium. The population, P, after t hours is given by the formula $P = (6 \times 10^6)(2.3)^t$. Find the population after four hours.

7. A radioactive material decays according to the formula $A = A_0\left(\frac{2}{3}\right)^t$, where A_0 is the initial amount present and t is measured in years. Find the amount present in five years.

8. The charge remaining in a battery decreases as the battery discharges. The charge, C (in coulombs), after t days is given by the formula $C = (3 \times 10^{-4})(0.7)^t$. Find the charge after five days.

9. Real estate in a particular city appreciates an average of 6% per year. If a house is valued at $85,000 today, what will it be worth in ten years?

10. A biologist makes a sample count of bacteria in a culture and finds that in 15 hours, it grows from 350 bacteria to 11,200 bacteria.
  (a) What is the doubling period of the bacteria?

  (b) How many bacteria are in the culture after three days?

11. A medical researcher requires a culture sample of 260,000 for an experiment to be conducted in 48 hours. If the tripling period for this culture is six hours, what size sample should she start with?

12. The half-life of a particular radioactive substance is eight years. Upon release into the atmosphere, the immediate area is not considered safe to inhabit until the amount of the substance is less than $\dfrac{1}{500}$ of its original amount. In approximately how many years will the area be safe again?

13. Each year, the value of a particular car depreciates by 20% of its previous year's value. If the initial value of the car was $23,500, how long will it take for the car to be worth only half of its original value?

14. Extend the table shown in Example 1 to show the effective annual interest rate if interest is compounded every day, every hour, every minute, and every second.

Properties of Logarithms

You have learned about logarithms and how they are related to exponents. The logarithm function is:

$$b^y = x \Leftrightarrow \log_b x = y$$

You used the logarithm function to investigate the following properties and laws of logarithms. Assume that $a, b, x, y, M, N > 0$, and $a \neq 1$ and $b \neq 1$.

1. $\log_b 1 = 0$ 4. $b^{\log_b x} = x$ 7. $\log_b M^N = N \cdot \log_b M$

2. $\log_b b = 1$ 5. $\log_b (M \cdot N) = \log_b M + \log_b N$ 8. if $\log_b x = \log_b y$, then $x = y$

3. $\log_b b^x = x$ 6. $\log_b (M \div N) = \log_b M - \log_b N$ 9. $\log_b x = \dfrac{\log_a x}{\log_a b}$

Example 1

Rewrite $5^{-2} = \dfrac{1}{25}$ as a logarithmic equation.

Solution

5 is the base and –2 is the exponent, or the value that the logarithm is equal to.

$$5^{-2} = \frac{1}{25}$$

$$5^{-2} = \frac{1}{25} \Leftrightarrow \log_5 \frac{1}{25} = -2$$

Example 2

Rewrite $\log_2 \dfrac{1}{8} = -3$ as an exponential equation.

Solution

2 is the base and –3 is the value of the logarithm, or the exponent.

$$\log_2 \frac{1}{8} = -3$$

$$\log_2 \frac{1}{8} = -3 \Leftrightarrow 2^{-3} = \frac{1}{8}$$

Example 3

Solve for x.

(a) $\log_4 1 = x$ **(b)** $\log_4 4 = x$ **(c)** $\log_4 4^7 = x$ **(d)** $7^{\log_7 16} = x$

Solution

(a) $\log_4 1 = x$ **(b)** $\log_4 4 = x$ **(c)** $\log_4 4^7 = x$ **(d)** $7^{\log_7 16} = x$

$\quad\quad 4^x = 1$ $\quad\quad 4^x = 4$ $\quad\quad 4^x = 4^7$ $\quad\quad \log_7 x = \log_7 16$

$\quad\quad\quad x = 0$ $\quad\quad\quad x = 1$ $\quad\quad\quad x = 7$ $\quad\quad\quad\quad x = 16$

Example 4

Rewrite $\log \frac{r}{st}$ in expanded form.

Solution

$$\log \frac{r}{st} = \log r - \log st \qquad \text{Quotient law of logarithms}$$
$$= \log r - (\log s + \log t) \qquad \text{Product law of logarithms}$$
$$= \log r - \log s - \log t \qquad \text{Distributive property}$$

Example 5

Evaluate $\log_3 14$.

Solution

$$\log_3 14 = \frac{\log 14}{\log 3} \qquad \text{Rewrite using the change-of-base formula.}$$
$$\approx 2.4022 \qquad \text{Evaluate using a calculator.}$$

Exercises

1. Explain in your own words.
    (a) the meaning of $\log_{10} 100 = 2$

    (b) why $b^{\log_b x}$ must be equal to x

    (c) why $\log_b b$ must be equal to 1

2. If the base of a logarithm is not explicitly written, as in log 5, what is the base assumed to be?

3. Use a calculator to verify each equation.
    (a) $\log [(2.5)(3.7)] = \log 2.5 + \log 3.7$

    (b) $\log \frac{11.3}{6.1} = \log 11.3 - \log 6.1$

    (c) $\log (2.25)^4 = 4 \log 2.25$

    (d) $\log 45.37 = \frac{\log 45.37}{\log 10}$

    (e) $\log \sqrt{24.3} = \frac{1}{2} \log 24.3$

4. Expand and simplify, but do not evaluate.
Assume that x, y, and z are positive numbers.

(a) $\log_2(4 \cdot 5)$

(b) $\log_3(27 \cdot 5)$

(c) $\log_6 \dfrac{x}{36}$

(d) $\log_8 \dfrac{y}{8}$

(e) $\log y^4$

(f) $\log z^9$

(g) $\log \sqrt{5}$

(h) $\log \sqrt[3]{7}$

5. Expand and simplify, but do not evaluate.

(a) $\log_3 27x$

(b) $\log \dfrac{100}{x}$

(c) $\log_5 \sqrt{27}$

(d) $\log 10ab$

6. Expand each expression. Assume that x, y, and z are positive numbers.

(a) $\log x^3 y^2$

(b) $\log 4xz$

(c) $\log x\sqrt{z}$

(d) $\log_3 \dfrac{x}{yz}$

(e) $\log_b \dfrac{x^2 y^3}{z}$

(f) $\sqrt{\log_8 \dfrac{x}{yz^2}}$

7. Write each expression as a single logarithm in condensed form. Assume that x, y, z, and b are positive numbers and $b \neq 1$.

(a) $\log_2(x+1) - \log_2 x$

(b) $\log_3 x - \log_3(x+2) - \log_3 8$

(c) $2\log x + \dfrac{1}{2}\log y$

(d) $\log_b \left(\dfrac{x}{z} + x \right) - \log_b \left(\dfrac{y}{z} + y \right)$

(e) $3\log_2 x - 5\log_2 y + 7\log_2 z$

(f) $-3 \log_b y - 7 \log_b z + \dfrac{1}{2}\log_b (x+2)$

8. Use the change-of-base formula to evaluate each logarithm to four decimal places.

   (a) $\log_5 9$ **(b)** $\log_5 4$

   (c) $\log_6 2$ **(d)** $\log_6 12$

   (e) $\log_7 5$ **(f)** $\log_7 83$

   (g) $\log_5 17$ **(h)** $\log_{17} 5$

9. If $\log_b 5 = 1.1609$ and $\log_b 8 = 1.5000$, find each value to four decimal places.

   (a) $\log_b 40$ **(b)** $\log_b 64$

   (c) $\log_b 25$ **(d)** $\log_b 1.6$

   (e) $\log_b 0.625$ **(f)** $\log_b 0.32$

10. Explain the difference between the logarithm of a product and the product of logarithms. Give an example.

11. Explain how to use the $\boxed{\text{log}}$ key on a calculator to find $\log_2 7$.

12. Evaluate.

   (a) $8^{\log_8 10}$ **(b)** $\log_5 5^2$

   (c) $\log_3 3^5$ **(d)** $\log_7 1$

   (e) $\log_3 3^7$ **(f)** $5^{\log_5 8}$

13. Find and correct the error in each solution.

   (a)
$$\begin{aligned}
\log \frac{x\sqrt[5]{y}}{z} &= \log x\sqrt[5]{y} - \log z \\
&= \log x \cdot \log \sqrt[5]{y} - \log z \\
&= \log x \cdot \log y^{\frac{1}{5}} - \log z \\
&= \log x \cdot \frac{1}{5}\log y - \log z
\end{aligned}$$

   (b)
$$\begin{aligned}
\log_{\sqrt{4}} \sqrt{7} &= \log \sqrt{7} - \log \sqrt{4} \\
&= \log 7^{\frac{1}{2}} - \log 4^{\frac{1}{2}} \\
&= \frac{1}{2}\log 7 - \frac{1}{2}\log 4 \\
&\approx 0.1215
\end{aligned}$$

   (c) If $\log_b 8 = 0.9031$, $\log_b 4 = 0.6021$, and $\log_b 12 = 1.0792$, evaluate $\log_b \dfrac{8}{48}$.

$$\begin{aligned}
\log_b \frac{8}{48} &= \log_b 8 - \log_b 48 \\
&= \log_b 8 - \log_b (4 \cdot 12) \\
&= \log_b 8 - (\log_b 4 + \log_b 12) \\
&= 0.9031 - (0.6021 + 1.0792)
\end{aligned}$$

12.5 Logarithmic Equations

In this lesson, you learned these steps for solving **exponential equations** using logarithms.

- Take the common logarithm of both sides.
- Use the property $\log M^x = x \log M$ to write an exponent as a coefficient.
- Manipulate the equation to solve for the variable.

You also learned how to solve **logarithmic equations** using a variety of methods including:

- Use the properties of logarithms to express each side of the equation as a single logarithm and then write the equation without the logarithms (if $\log M = \log N$, then $M = N$).
- Isolate a constant on one side, express the other side as a single logarithm, and then change to equivalent exponential form to solve for the variable.

Example 1

Solve the equation: $6^{x-3} = 2^x$

Solution

$$\log 6^{x-3} = \log 2^x \qquad \text{Take the common logarithm of both sides.}$$
$$(x-3)\log 6 = x \log 2 \qquad \log M^x = x \log M$$
$$x \log 6 - 3 \log 6 = x \log 2 \qquad \text{Apply the distributive property.}$$
$$x \log 6 - x \log 2 = 3 \log 6 \qquad \text{Collect terms with } x \text{ on one side.}$$
$$x(\log 6 - \log 2) = 3 \log 6 \qquad \text{Factor out the common factor.}$$
$$x = \frac{3 \log 6}{\log 6 - \log 2} \qquad \text{Isolate } x.$$
$$x \approx 4.89279 \qquad \text{Use a calculator.}$$

Example 2

Solve the equation: $\log x = 1 - \log (x-3)$

Solution

$$\log x + \log (x-3) = 1 \qquad \text{Isolate the constant.}$$
$$\log x(x-3) = 1 \qquad \text{Express the left side as single logarithm.}$$
$$10^1 = x(x-3) \qquad \log_b y = a \text{ means } b^a = y. \text{ Change to exponential form.}$$
$$10 = x^2 - 3x \qquad \text{Apply the distributive property.}$$
$$0 = x^2 - 3x - 10$$
$$0 = (x-5)(x+2) \qquad \text{Factor.}$$
$$x = 5 \text{ or } x = -2$$

Check $x = 5$

L.S.　R.S.

$\log 5$　$1 - \log (5-3)$
$= \log 10 - \log 2$
$= \log \frac{10}{2}$
$= \log 5$

L.S. = R.S., so $x = 5$ is a solution.

Check $x = -2$

L.S.	R.S.

$\log\ (-2)$ $1 - \log\ (-2 - 3)$

$= \log\ 10 - \log\ (-5)$

$= \log\ \dfrac{10}{-5}$

$= \log(-2)$

Since it's not possible to take the logarithm of a negative number, $x = -2$ is not a solution.

Example 3

Solve the equation: $2\log_5\ n = \log_5\ 8 + \log_5\ (n - 2)$

Solution

$$2\log_5\ n = \log_5\ 8 + \log_5\ (n - 2)$$
$$\log_5\ n^2 = \log_5\ 8 + \log_5(n - 2) \qquad \log\ M^x = x\log\ M$$
$$\log_5\ n^2 = \log_5\ 8(n - 2) \qquad \text{Express the right side as a single logarithm.}$$
$$n^2 = 8(n - 2) \qquad \log\ M = \log\ N \text{ thus } M = N$$
$$n^2 = 8n - 16 \qquad \text{Apply the distributive property.}$$
$$n^2 - 8n + 16 = 0 \qquad \text{Write as a quadratic equation.}$$
$$(n - 4)(n - 4) = 0 \qquad \text{Factor to solve.}$$
$$n = 4$$

Check $n = 4$

L.S.	R.S.
$2\log_5\ 4$	$\log_5\ 8 + \log_5\ (4 - 2)$
$= \log_5\ 4^2$	$= \log_5\ 8 + \log_5\ 2$
$= \log_5\ 16$	$= \log_5\ 16$

L.S. = R.S., so the solution is $n = 4$.

Exercises

1. Complete the sentences.
   (a) The expression $\log_3\ (4x)$ is the logarithm of a _____.

   (b) The expression $\log_2 \dfrac{5}{x}$ is the logarithm of a _____.

   (c) The expression $\log\ 4^x$ is the logarithm of a _____.

   (d) In the expression $\log_5\ 4$, the number 5 is the _____ of the logarithm.

   (e) An equation with a variable in its exponent, such as $3^{2x} = 8$, is called an _____ equation.

   (f) An equation with a logarithmic expression that contains a variable, such as $\log_5(2x - 3) = \log_5(x + 4)$, is a _____ equation.

2. Explain how to solve each equation, and then solve it.
   (a) $3^x = 7$ (b) $2^{x + 1} = 32$ (c) $7^{x^2} = 10$

3. Fill in the blanks to complete each solution.

(a)
$$2^x = 7$$
$$\boxed{}\, 2^x = \log 7$$
$$x\,\boxed{} = \log 7$$
$$x = \frac{\log 7}{\log 2}$$

(b) $\log_2 (2x - 3) = \log_2 (x + 4)$
$$\boxed{} = x + 4$$
$$x = 7$$

4. Solve.

(a) $\log 2a = \log 4$

(b) $\log (3 - 2x) - \log (x + 24) = 0$

(c) $\dfrac{\log (8y - 7)}{\log y} = 2$

5. **(a)** Check to see whether $x = -4$ is a solution of $\log_5 (x + 3) = \dfrac{1}{5}$.

(b) Check to see whether $w = -2$ is a solution of $5^{2w+3} = \dfrac{1}{5}$.

6. Verify the identity: $2 \log \sqrt{x} = \log x$

7. Solve.

(a) $\log x^3 = 3$

(b) $\log (8y + 12) = 2$

(c) $\log \dfrac{4a + 1}{2a + 9} = 0$

(d) $\log_b 5 = \dfrac{1}{4}$

8. Solve each equation using two different methods.
(a) $2 \log (y + 2) = \log (y + 2) - \log 12$

(b) $\log_6 (x + 3) + \log_6 (x - 2) = 1$

9. Solve.

(a) $13^{a-1} = 2$

(b) $3^{2b} = 4^b$

(c) $2^{x+1} = 7$

(d) $5^{y-3} = 3^{2y}$

(e) $5^{x^2} = 2^{5x}$

(f) $8^{a^2} = 11$

10. Solve. Remember to check the solutions.

(a) $\log_5 (7x + 1) - \log_5 (x - 1) = 2$

(b) $\log_8 (x - 4) = 1 - \log_8 (x + 3)$

(c) $1 - \log (w - 4) = \log (w + 5)$

(d) $\log a + \log (a + 9) = 1$

(e) $\log_7 (2x + 2) - \log_7 (x - 1) = \log_7 (x + 1)$

(f) $\log_3 w = \log_3 \dfrac{1}{w} + 4$

(g) $2 \log_3 x - \log_3 (x - 4) = 2 + \log_3 2$

(h) $\log_8 x - \log_8 (x - 1) = 1$

(i) $\log_3 (2w - 8) - \log_3 (3w + 2) = 1$

11. Eric solved the equation $3^{x+2} = 6^x$, but when he checked his answer, it was incorrect. Find and correct the error in Eric's solution.

$$3^{x+2} = 6^x$$
$$\log 3^{x+2} = \log 6^x$$
$$x + 2 \log 3 = x \log 6$$
$$x - x \log 6 = -2 \log 3$$
$$x(1 - \log 6) = -2 \log 6$$
$$x = \frac{-2 \log 6}{(1 - \log 6)}$$
$$x \approx 3.2851$$

12.6 Graphing Logarithmic Functions

- The exponential function $y = b^x$ and the logarithmic function $y = \log_b x$ are inverses of each other, so they are symmetric about the line $y = x$. The function $y = \log_b x$ has these properties:
- When $b > 1$, the function is increasing, and when $0 < b < 1$, the function is decreasing.
- The graph of the function passes through the points $(1, 0)$ and $(b, 1)$.
- The line $x = 0$ is a vertical asymptote.
- The domain is $\{x \mid x > 0, x \in \mathrm{R}\}$. The range is $\{y \mid y \in \mathrm{R}\}$.
- The general form of a vertical translation of $y = \log_b x$ is $y = \log_b x + k$. The translation is k units up if k is positive, and k units down if k is negative. The asymptote, domain, and range of the vertically translated function are the same as those of the original function.
- The general form of a horizontal translation of $y = \log_b x$ is $y = \log_b (x + d)$. The translation is d units to the left if $d > 0$, and d units to the right if $d < 0$. The vertical asymptote is the line $x = -d$. The range of the translated function is the same as the original range, and the domain changes to $\{x \mid x > -d, x \in \mathrm{R}\}$.

Example 1

Draw the graph of $y = \log_3 x$.

Solution

Method 1: Use a table of values.

Change $y = \log_3 x$ to exponential form and choose values for y to get corresponding values for x.

$$y = \log_3 x \Leftrightarrow x = 3^y$$

x	y
$\frac{1}{3}$	–1
1	0
3	1
9	2
27	3
81	4

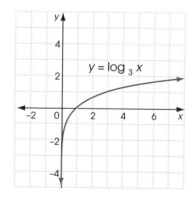

Method 2: Use a graphing calculator.

Use the change-of-base formula to convert the logarithm to a common logarithm, then use a graphing calculator.

$$y = \log_3 x \Leftrightarrow y = \frac{\log x}{\log 3}$$

Example 2

Graph the function $y = \log_2 x$. Then graph $y = \log_2 (x + 3) - 4$ on the same grid.

Solution

The general form of a vertical translation of $y = \log_b x$ is $y = \log_b x + k$. For the function given, $k = -4$, so translate the graph 4 units down.

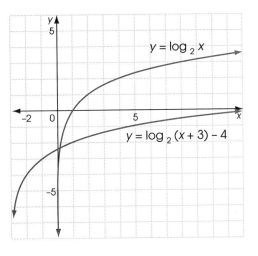

The general form of a horizontal translation of $y = \log_b x$ is $y = \log_b (x + d)$. For the function given, $d = 3$, so translate the graph 3 units to the left.

The graph $y = \log_2 (x + 3) - 4$ is identical to the graph of $y = \log_2 x$, except that it is translated 3 units to the left and 4 units down.

Example 3

Analyze the graph of $y = \log_{10} (2x + 1)$. Identify:

(a) the domain and the range.

(b) the equation of the asymptote.

(c) the x-intercept.

(d) the x-coordinate when $y = 1$.

Solution

First draw the graph using a table of values or a graphing calculator.

(a) The domain is $\left\{ x \mid x > -\dfrac{1}{2},\ x \in R \right\}$.

The range is $\{ y \mid y \in R \}$.

(b) The equation of the asymptote is $x = -\dfrac{1}{2}$.

(c) To find the x-intercept, substitute $y = 0$ into the equation.
$$0 = \log_{10} (2x + 1)$$
$$2x + 1 = 1$$
$$x = 0$$

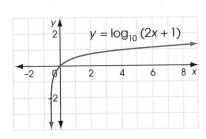

(d) Substitute $y = 1$ into the equation.
$$1 = \log_{10} (2x + 1)$$
$$2x + 1 = 10$$
$$x = \dfrac{9}{2}$$

Exercises

1. Explain the relationship between $y = \log x$ and $y = 10^x$.

2. Explain why the domain of the logarithmic function $y = \log_b x$ does not include negative numbers.

3. Graph each equation and describe how it compares to the graph of $y = \log x$.
   (a) $y = \log (x + 2)$ (b) $y = -\log x$ (c) $y = \log x - 2$

4. Tell whether each statement is true or false. Correct each false statement.
   (a) $y = \log_2 x$ is a logarithmic function with a base of 2.

   (b) $(0, 1)$ is a point common to all logarithmic functions.

   (c) $y = \log_b x$ is equivalent to $b^y = x$.

   (d) If $x = 0$, then $\log_b x = 0$.

   (e) $y = \log_b x$ is not defined for $x < 0$.

   (f) The inverse of $\log_5 x = y$ could be written as $\log_5 y = x$.

5. The graph of $f(x) = \log_4 x$ is shown.
   (a) What is the domain of the function?

   (b) What is the range of the function?

   (c) What is the y-intercept of the graph?

   (d) What is the x-intercept of the graph?

   (e) Is the function increasing or decreasing?

   (f) The graph passes through the point $(4, y)$. What is y?

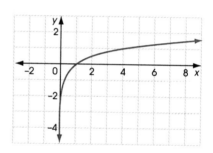

6. The graph of $y = \log_{10} x$ is shown.
   (a) Find the approximate value of $\log_{10} 4$.

   (b) Find the approximate value of $\log_{10} 8$.

   (c) How are $y = \log_{10} x$ and $10^y = x$ related?

   (d) If y is 0.15, find the approximate value of x.

   (e) If y is 0.75, find the approximate value of x.

   (f) Use your calculator to find the values of x when $y = 0.15$ and when $y = 0.75$. Compare your calculated values to the approximate values obtained by using the graph. (Hint: Enter the exponential forms, $10^{0.15}$ and $10^{0.75}$, in your calculator.)

7. If point $P (a, b^a)$ is a point on the graph of $y = b^x$, find the coordinates of a corresponding point Q on the graph of $y = \log_b x$.

8. Complete the table of values. If an evaluation is not possible, write "none."

(a)

$f(x) = \log_5 x$	
x	**y**
$\dfrac{1}{25}$	
$\dfrac{1}{5}$	
1	
5	
25	

(b)

$f(x) = \log_6 x$	
x	**y**
-6	
0	
$\dfrac{1}{216}$	
$\sqrt{6}$	
6^8	

(c)

$f(x) = \log_8 x$	
x	**y**
-8	
0	
$\dfrac{1}{8}$	
$\sqrt{8}$	
64	

9. Draw each pair of inverse functions on the same grid.
Draw and label the axis of symmetry.

(a) $y = 6^x$ and $y = \log_6 x$ **(b)** $y = 3^x$ and $y = \log_3 x$

(c) $y = 5^x$ and $y = \log_5 x$ **(d)** $y = 8^x$ and $y = \log_8 x$

10. Draw the graphs of $y = \left(\dfrac{1}{3}\right)^x$ and its inverse on the same grid.
Draw and label the axis of symmetry.

11. Graph each function and tell whether it is increasing or decreasing.
(a) $y = \log_3 x$

(b) $y = \log_{\frac{1}{3}} x$

(c) $y = \log_{\frac{1}{2}} x$

(d) $y = \log_4 x$

12. Graph each function without using a table of values.

(a) $y = 3 + \log_3 x$ **(b)** $y = \log_{\frac{1}{3}}\left(x - 2\right)$

(c) $y = \log_{\frac{1}{2}}(x - 1)$ **(d)** $y = \log_4 (x + 2)$

13. List as many methods as you can to solve the equation $3^x = 12$. Explain each
method completely. Which method do you prefer? Justify your preference.

Applications of Logarithms

In this lesson, you solved a variety of problems using exponents and logarithms.

- To solve an exponential equation that has a variable in the exponent, first isolate the exponent by taking the logarithm of both sides of the equation. Apply the power property of logarithms to convert the exponent into a coefficient, and then isolate the variable.

- To solve an exponential equation that has a constant in the exponent, first determine the value of the constant, then evaluate the final answer.

Example 1

An initial population of 200 bacteria increased to 1,200 in 4 hours. How long will it take for the population to reach 10,000?

Solution

Use the population growth formula: $P = P_0 2^{kt}$

P = population at time t
P_0 = initial population
k = population constant
t = time

Substitute $P_0 = 200$, $P = 1,200$, and $t = 4$ into the formula, and solve for k.

$$1,200 = 200 \cdot 2^{k(4)}$$
$$6 = 2^{4k}$$
$$\log 6 = 4k \log 2$$
$$k = \frac{\log 6}{4 \cdot \log 2}$$

Now substitute $P_0 = 200$, $P = 10,000$, and $k = \frac{\log 6}{4 \cdot \log 2}$ into the formula, and solve for t.

$$10,000 = 200 \cdot 2^{\frac{\log 6}{4 \cdot \log 2}t}$$
$$50 = 2^{\frac{\log 6}{4 \cdot \log 2}t}$$
$$\log 50 = \log 2^{\frac{\log 6}{4 \cdot \log 2}t}$$
$$\log 50 = \left(\frac{\log 6}{4 \cdot \log 2}\right)t \cdot \log 2$$
$$t = \frac{4 \cdot \log 50}{\log 6}$$
$$t \approx 8.7$$

It will take approximately 8.7 hours for the bacteria to increase to a population of 10,000.

Example 2

If $800 is deposited in an account paying 4.5% interest, compounded semiannually, how long will it take for the account to increase to $2,000?

Solution

Use the compound interest formula: $A = P(1 + i)^n$

A = amount
P = principal
i = interest rate per compounding period
$i = \dfrac{0.045}{2}$
n = number of compounding periods

Substitute $A = 2,000$, $P = 800$, and $i = \dfrac{0.045}{2}$ into the formula, and solve for n.

$$2,000 = 800\left(1 + \frac{0.045}{2}\right)^n$$
$$2.5 = (1 + 0.0225)^n$$
$$2.5 = (1.0225)^n$$
$$\log 2.5 = n \log 1.0225$$
$$n = \frac{\log 2.5}{\log 1.0225}$$
$$n \approx 41.2$$

It will take 42 compounding periods, or 21 years, for the account to increase to $2,000.

Exercises

1. Explain, in your own words, the relationship between logarithms and exponents.

2. Use the Richter scale formula to calculate the magnitude of each earthquake.

   $R = \log \dfrac{A}{P}$ where R = magnitude, A = amplitude, P = period

   (a) amplitude = 4,000 micrometers, period = 0.1 seconds

   (b) amplitude = 2,500 micrometers, period = 0.25 seconds

   (c) amplitude = 6,000 micrometers, period = 0.2 seconds

   (d) amplitude = 3,500 micrometers, period = 0.2 seconds

3. Use the formula for dB gain to calculate the decibel voltage gain of each amplifier.

dB gain $= 20 \log \dfrac{E_O}{E_1}$ where E_O = output voltage, E_1 = input voltage

(a) input voltage = 0.5 volts, output voltage = 20 volts

(b) input voltage = 0.25 volts, output voltage = 25 volts

(c) input voltage = 0.2 volts, output voltage = 40 volts

(d) input voltage = 0.3 volts, output voltage = 64 volts

4. An earthquake measuring 4.7 on the Richter scale has a period of 0.1 seconds. Find the amplitude of the earthquake.

5. An amplifier with an input voltage of 0.5 volts provides a 40 decibel voltage gain. Find the output voltage.

6. How old is a wooden statue that has only one-fourth of its original carbon-14 content? Use the formula given, and assume that carbon-14 has a half-life of about 5,700 years.

$A = A_0 \cdot 2^{-\frac{t}{h}}$ where A = amount of material at time t,
A_0 = original amount of material (at $t = 0$), and h = half-life

7. If $600 is put into an account paying 6.5% interest, compounded quarterly, how long will it take for the account to increase to $1,000?

8. The population of a town is expected to double every 30 years. How long will it take for the present population of 70,000 to grow to 100,000?

9. In 1999, Uganda was struck by an earthquake measuring 4.1 on the Richter scale. Ten years earlier, San Francisco was struck by an earthquake measuring 7.1. Assume the periods were the same for both earthquakes and calculate how many times larger the amplitude of the San Francisco earthquake was.

10. In three years, 25% of a radioactive element decays. Find its half-life.

11. If $500 is put into an account paying 7.5% interest, compounded monthly, how long will it take for the account to increase to $2,000?

12. Lead-201 has a half-life of 8.4 hours. How many hours ago was there 30% more of the substance?

13. Create a problem involving the half-lives of some of these isotopes:

Isotope	Half-life
argon-42	33 years
potassium-48	69 seconds
calcium-51	10 seconds
cobalt-57	271 days
copper-69	3 minutes
carbon-14	5,730 years

12.8 Base-*e* Exponential Functions

The natural exponential function is $f(x) = e^x$ where e is the value of $\left(1 + \frac{1}{n}\right)^n$ as n increases to ∞. e is an irrational number approximately equal to 2.718281828409. Some characteristics of the natural exponential function are:

- It is an increasing function.
- The domain is the set of real numbers, $(-\infty, \infty)$.
- The range is the set of positive real numbers, $(0, \infty)$.
- The graph of $y = e^x$ passes through $(0, 1)$ and $(1, e)$.
- The asymptote is the *x*-axis, which has the equation $y = 0$.
- A horizontal translation of the graph by c units has an equation of the form $f(x) = e^{(x + c)}$; the translation is left if c is positive, right if c is negative.
- A vertical translation of the graph by c units has an equation of the form $f(x) = e^x + c$; the translation is up if c is positive, down if c is negative.

The formula for continuous exponential growth is modeled by $A = Pe^{rt}$, where P is the initial amount, r is the rate of increase, and t is the time. Continuous exponential decay is modeled with the same formula, where r is negative.

Example 1

What transformations must be performed on the graph of $y = e^x$ to yield the graph of $y = -e^{x-1}$? Sketch both graphs on the same coordinate grid.

Solution

The negative sign of the lead factor in $y = -e^{x-1}$ indicates a reflection in the *x*-axis, and $x - 1$ in the exponent represents a horizontal translation of 1 unit to the right. The original and translated graphs are shown.

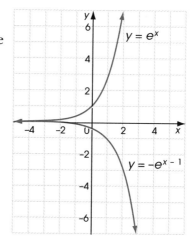

Example 2

$10,000 is invested for 1 year at an interest rate of 9% per year. Compare the yields with monthly compounding and continuous compounding.

Solution

Compound interest

Compound interest formula: $A = P\left(1 + \frac{r}{k}\right)^{kt}$

$$A = 10,000\left(1+\frac{0.09}{12}\right)^{(12)(1)}$$
$$= 10,000(1.0075)^{12}$$
$$\approx 10,938.07$$

Continuous interest

Continuous interest formula: $A = Pe^{rt}$

$$A = 10,000e^{(0.09)(1)}$$
$$\approx 10,941.74$$

Continuous compounding would yield $3.67 more.

Example 3

The downward velocity of a skydiver in free fall is given by the formula $v(t) = 50(1 - e^{-0.2t})$, where v is the velocity in meters per second after t seconds. Determine the initial velocity and the velocity after 10 seconds.

Solution

The initial velocity occurs when $t = 0$.

$$v(0) = 50(1 - e^{(-0.2)(0)})$$
$$= 50(1 - e^{0})$$
$$= 50(1 - 1)$$
$$= 0$$

The initial downward velocity is 0 m/s.

After 10 seconds, the velocity will be

$$v(10) = 50(1 - e^{(-0.2)(10)})$$
$$= 50(1 - e^{-2})$$
$$\approx 43$$

The velocity after 10 seconds is 43 m/s.

Exercises

1. A reporter wrote that "the sport of snowboarding is growing exponentially." In your own words, write a paragraph to explain what this means.

2. Graph the function $f(x) = e^x$.

   (a) What special name does the function $f(x) = e^x$ have?

   (b) What is the x-intercept of the function $f(x) = e^x$?

3. Evaluate each expression.
   (a) e^4

   (b) $e^{-3.3}$

   (c) $e^5 \cdot e^{-2}$

   (d) $6e^{10} \div 2e^4$

   (e) $(2e^{64})^{\frac{1}{8}}$

4. Is $T(t) = 68 + 220e^{-0.2t}$ an increasing or a decreasing function?

5. Plot the numbers on a number line.
$$\left\{ \pi, \ e, \ \sqrt{2}, \ \frac{\sqrt{3}}{2} \right\}$$

6. Identify the transformation(s) of the graph of $y = e^x$ that would yield the graph of each function, and then sketch the graph.
   (a) $y = e^x - 2$ (b) $y = e^{x+3}$ (c) $y = e^{-x} - 1$

7. An initial investment of \$5,000 earns 8.2% interest, compounded continuously. What will the investment be worth in 12 years?

8. An initial investment of \$5,000 grows at an annual rate of 8.5% for 5 years. Compare the final balances that result from annual compounding and continuous compounding.

9. An account now contains \$3,610 and has been accumulating interest at an annual rate of 8% compounded continuously. How much was in the account 4 years ago?

10. The population of Earth is approximately 6 billion people and is growing at an annual rate of 1.9%. Assuming growth is modeled by $A = Pe^{rt}$, find the world population in 30 years.

11. The spread of a disease through a herd of cattle can be modeled by the formula $A = Pe^{0.27t}$, where t is measured in days. If a herd has 2 infected animals and they remain untreated, how many cattle will have the disease in 12 days?

12. The population of a country is increasing at the rate of 3% annually and is modeled by the formula $A(t) = 10,000e^{0.03 \cdot t}$. The number of people that can be fed by the country's food supply is modeled by the equation $F(t) = 40.5t + 14,000$. How long will it take for the population to outstrip the food supply?

13. A student used $A = \$3,610e^{(0.08)(-4)}$ to solve Problem 9. Discuss with a partner the problems with this formula and why the student might have written it this way.

14. Find and correct the solution to the given problem.

Problem

The growth of bacteria in a culture is modeled by the formula $N = N_0 e^{rt}$, where N_0 is the original number of bacteria, r is the growth rate (expressed as a decimal), and t is the time in minutes. If originally there were 2,500 bacteria, and the number is increasing by 2.5% each minute, how many bacteria are there after an hour and a half?

$$N = 2,500e^{(0.025)(1.5)}$$
$$= 2,500e^{22.5}$$
$$\approx 2,596$$

Solution

After an hour and a half, there were about 2,596 bacteria in the culture.

12.9 Base-*e* Logarithms

In this lesson, you used your knowledge of logarithms and base-*e* exponential functions to study the natural logarithm.

- $y = \log_e x = \ln x$
- $x = \ln y$ is equivalent to $e^x = y$
- The natural exponential function, $y = e^x$, and the natural logarithm, $y = \ln x$, are inverse functions.
- Natural logarithms have the same properties as other logarithms.
- Graphs of equations of the general form $f(x) + a$ are vertical translations of the graph of $y = \ln x$. The graph of $y = \ln x$ is translated a units up if a is positive and a units down if a is negative.
- Graphs of equations of the general form $f(x + a)$ are horizontal translations of the graph of $y = \ln x$. The graph of $y = \ln x$ is translated a units to the left if a is positive and a units to the right if a is negative.

Example 1

The nuclear disaster at Chernobyl in 1986 released approximately 13,000 kg of radioactive iodine into the atmosphere. The amount of radioactive iodine, A, left after t days is given by the formula

$$A(t) = 13{,}000e^{\frac{-t}{11.6}}$$

(a) Calculate the amount of radioactive iodine remaining after 30 days.

(b) How long will it take for the amount of radioactive iodine to be reduced to one-half of its original amount.

Solution

(a) Substitute $t = 30$ into the formula.

$$A(30) = 13{,}000e^{\frac{-30}{11.6}}$$
$$= 979 kg$$

979 kg of radioactive iodine remain after 30 days.

(b)

$\quad 6{,}500 = 13{,}000e^{\frac{-t}{11.6}}$ — Substitute one-half of 13,000 kg into the formula for A. $\frac{1}{2}$ of 13,000 is 6,500.

$\quad 0.5 = e^{\frac{-t}{11.6}}$ — Divide both sides of the equation by 13,000.

$\quad \ln 0.5 = \frac{-t}{11.6}$ — Write the equivalent logarithmic equation.

$\quad 11.6 \ln 0.5 = -t$ — Multiply both sides by 11.6.

$\quad 11.6 \ln 0.5 = t$ — Multiply both sides by –1.

Use a calculator.

$$t \approx 8 \text{ days (rounded)}$$

It would take about 8 days for the radioactive iodine to be reduced to one-half of its original amount.

Example 2

Complete the table. Refer to the graph.

Graph Number	Vertical/Horizontal Translation of $y = \ln x$	Amount of the Translation	Equation
1			
2			

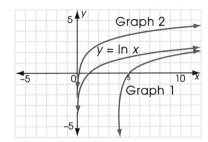

Solution

Graph Number	Vertical/Horizontal Translation of $y = \ln x$	Amount of the Translation	Equation
1	horizontal	4 unts right	$y = \ln(x - 4)$
2	vertical	2 units up	$y = \ln x + 2$

Example 3

Write the expression as a single logarithm.

$$\ln a + \ln b - \frac{1}{2} \ln c$$

Solution

$$\ln a + \ln b - \frac{1}{2} \ln c = \ln ab - \frac{1}{2} \ln c$$
$$= \ln ab - \ln c^{\frac{1}{2}}$$
$$= \ln \frac{ab}{c^{\frac{1}{2}}}$$

Exercises

1. Explain the differences between $y = \log x$ and $y = \ln x$.

2. How are $y = \ln x$ and $y = e^x$ related?

3. Fill in the blanks to make the statements true.
   (a) $y = \ln x$ and $y = e^x$ are _____ functions.
   (b) The graph of $y = \ln x$ has _____ as an asymptote.
   (c) The domain of the function $y = \ln x$ is the interval _____.
   (d) The graph of $y = \ln x$ passes through the point (__ , 0).
   (e) The statement $y = \ln x$ is equivalent to the exponential statement _____.
   (f) The logarithm of a negative number is _____.

4. Graph $y = e^x$, $y = \ln x$, and $y = x$. Explain the connections between the three graphs.

5. The function $y = \dfrac{1}{x}$ is its own inverse. Explain what this means.

6. Graph and describe the connection between each graph and the graph of $y = \ln x$.
   (a) $y = \ln x - 2$ **(b)** $y = \ln x + 2$

   (c) $y = \ln (x - 2)$ **(d)** $y = \ln (x + 2)$

7. Use the formula $t = \dfrac{\ln 2}{r}$ to find the doubling time for a population that is growing 5% per year.

8. Use the formula $r = \dfrac{\ln 2}{t}$ to find the growth rate, r, when the doubling time is 15 years.

9. Solve the following equations by switching forms.
   (a) $2 = \ln x - 1$

   (b) $5 = 2e^{x-1} + 2$

10. Write as a single logarithm: $\ln s - \ln f - 3 \ln k$

11. Expand: $\ln \dfrac{x^2 y^3}{z^5}$

12. Rosa makes Jell-O by combining the contents of a package with boiling water, then placing the mixture in a refrigerator. The time, t (in hours), that it takes the Jell-O to reach a temperature of T is given by the formula $t = 1.1 \ln \dfrac{9T + 140}{590 - 9T}$. Estimate the time it will take for the Jell-O to cool to 50°F.

13. Write the formula $A = Pe^{rt}$ in logarithmic form. Explain why it is useful to have the formula in each form.

14. Fill in the table of values, where t is the doubling time of the population and r is the rate of growth of the population.

r	1%	2%	3%	4%	5%	6%	7%	8%
t								

Draw a graph of t versus r. Use this graph to predict the doubling time at a growth rate of 7.5%.

15. Show the growth of $1,000 over 10 years at each rate.
   (a) 8%, compounded annually

   (b) 8%, compounded semiannually

   (c) 8%, compounded monthly

   (d) 8%, compounded continuously

Answers

1.1 Data in Tables I

1. A trend is a pattern in data that you can use to compare the present to the past or to make a prediction for the future.

3. (a) *examples:* (i) No one store is cheapest for everything. (ii) The price of tomato sauce is between 25¢ and 39¢ at every store. (iii) Price Way and Food Plus have higher prices for more items than the other store.
   (b) Snack Mart for milk, Price Way for the complete list
   (c) $P = 3g + 2t + 4b$
   (d) Snack Mart, at $14.11

5. (a) Brooks 41, Clark 35, Jao 45, Nichols 51, Pierce 45, Tyler 47
   (b) 44
   (c) $P = wh + 0.5w(h - 40)$, where P is gross pay, w is the employee's hourly wage, and h is the total number of hours worked

1.2 Data in Tables II

1. If the table is recursive, information in one row depends on information in a previous row.

3. $interest\ rate = \dfrac{interest}{principal}$. Round the result to a decimal in hundredths, then express the number of hundredths as a percent.

5. (a) 0.5%   (b) $55.76
   (c) *Interest Earned* $= 0.005 \cdot (O + 400)$
   (d) $5,430.45

7. (a) After the first row of data, the opening balance is equal to the year-end balance from the year before.

1.3 The Real Number System

1. natural, whole, integer, rational, real

3.

5. (a) (ii)   (b) (iv)   (c) (vi)
   (d) (v)   (e) (i)   (f) (iii)

7. *examples:*
   (a) rational, because the number is natural

(b) irrational, because $\sqrt{5}$ is a nonterminating, nonrepeating decimal

(c) irrational, because π is irrational, so 2π must also be irrational

(d) rational, because 9 is a natural number

(e) rational, because 0.1875 terminates

(f) rational, because $15.125 = 15\frac{1}{8}$ or $\frac{121}{8}$

(g) irrational, because the number is a nonterminating, nonrepeating decimal

(h) rational, because the decimal has a repeating period

(i) irrational, because the number is a nonterminating, nonrepeating decimal

9.

Number	Square Root
0.01	0.1
0.02	0.141421356...
0.03	0.17320508...
0.04	0.2
0.05	0.223606797...
0.06	0.244948974...
0.07	0.264575131...
0.08	0.282842712...
0.09	0.3
0.10	0.316227766...

Note: Although $\sqrt{0.03}$ may appear to terminate after 8 digits, this only happens because the next digit is a 0. Relate $\sqrt{\dfrac{3}{100}}$ to $\sqrt{3}$. There is at least one more digit, 8, after the 0.

(a) 0.01, 0.04, 0.09
(b) 0.02, 0.03, 0.05, 0.06, 0.07, 0.08, 0.1

11. (d) $\sqrt{2}$
   (e) Measure the diagonal with a string or paper strip and mark a distance exactly that far from 0 along the number line.
   (f) $\sqrt{3} = 1.732050808$, $\sqrt{5} = 2.236067978$, $\sqrt{7} = 2.645751311$

1.4 Solving Problems

1. *examples:* +: sum, added to, plus, more than, increased by; −: difference, subtracted from, minus, less than, decreased by; ·: multiplied by, increased, of, squared; ÷: divided by, quotient, shared into, groups of, ratio

3. (d) and (f) are both equal to $\dfrac{6^3}{4}$ or 54

5. (a) $\sqrt{6} = 2.45$ **(b)** $10 - 6 = 4$
(c) $12^2 = 144$ **(d)** $17 + 13 = 30$
(e) $\dfrac{1}{4}$ **(f)** $1 + 2 + 3 + 4 + 5 + 6 = 21$
(g) $\sqrt{243} = 15.59$ **(h)** $9^2 + 26 = 107$
(i) $\dfrac{62 + 73}{2} = 67.5$ **(j)** $0.2 \cdot 600 = 120$
(k) $\dfrac{1}{6} + \dfrac{1}{9} + \dfrac{1}{12} = \dfrac{13}{36}$

7. Sachi multiplied before subtracting, but the problem says to subtract first, then find a percent of the result. Also, the decimal form of 40% is 0.4, not 40.
Corrected solution:
$(12^3 - 5^2) \cdot 40\% = (1{,}728 - 25) \cdot 0.4$
$= 1{,}703 \cdot 0.4$
$= 681.2$

9. To find the reciprocal of the square root of twenty-five, enter: ⊡1⊡ ⊡÷⊡ ⊡2⊡ ⊡5⊡ ⊡√⊡

1.5 Operations with Real Numbers

1. *example:* To find $\sqrt{12}$:
Number Line: $\sqrt{12}$ is between perfect squares $\sqrt{9}$ and $\sqrt{16}$, so it is between 3 and 4. Since 12 is a bit less than halfway between 9 and 16, estimate about 3.4.
Fractions: $\sqrt{12}$ is between perfect squares $\sqrt{9}$ and $\sqrt{16}$, so it is between 3 and 4. The distance from 9 to 16 is 7 units, and the distance from 9 to 12 is 3 units, so $\sqrt{12}$ must be $\dfrac{3}{7}$ more than $\sqrt{9}$. $\sqrt{12}$ is about $3\dfrac{3}{7}$ or 3.428571428.
A calculator gives 3.464101615 as an approximation for $\sqrt{12}$, so using fractions gives a much more accurate result.

3. *examples:*
  (i) Calculator shows $\sqrt{15}$ is about 3.8729833346.
  (ii) Since you need to round to three decimal places, look at the number in the fourth decimal place, 9.
  (iii) Since 9 is greater than 4, round the third decimal place up from 2 to 3. The rounded value is 3.873.

5. *example:* ⊡3⊡ ⊡2ndF⊡ ⊡√⊡ ⊡1⊡ ⊡5⊡ ⊡=⊡

7. *examples:*
  (i) $2 \cdot \sqrt{3} \cdot 4 \cdot \sqrt{3} =$ **(ii)** $+(3 \cdot \sqrt{5}) =$
  (iii) $-(2 \cdot \sqrt{2}) =$ **(iv)** $\div(5 \cdot \sqrt{2}) =$
  (v) Round to two decimal places. Result is 3.94.

9. (i) Dane's initial quotient has a digit missing. It should be 2.2̲24859546.

(ii) In the second step, Dane should have multiplied before he added. The correct solution is:

$$\dfrac{3\sqrt{11}}{2\sqrt{5}} \approx \dfrac{9.949874371}{4.472135955}$$
$$\approx 2.224859546$$
$$2.224859546 \cdot \sqrt{6} \approx 5.449770637$$
$$5.449770637 + 4\sqrt{2} \approx 11.10662489$$
$$\approx 11.11$$

11. 57 ft

2.1 Rational Exponents

1. root index is 7; radicand is 9^3

3. (a) $x^{\frac{1}{3}}$ **(b)** $3^{-\frac{2}{5}}$ **(c)** $94^{\frac{17}{6}}$
  (d) $\dfrac{1}{52^{\frac{7}{2}}}$ **(e)** $(13x)^{\frac{2}{3}}$

5. $x \geq 0$, since a negative number cannot be multiplied by itself an even number of times to give a negative result. (*example:* $\sqrt{9}$ is a real number, but $\sqrt{-9}$ is not.)

7. (a) -9 **(b)** 125 **(c)** 16 **(d)** 1,024 **(e)** 4

9. $\sqrt[3]{\dfrac{1}{x^2}} \cdot \sqrt{x^3} = \sqrt[3]{x^{-2}} \cdot \sqrt{x^3}$
$= x^{-\frac{2}{3}} \cdot x^{\frac{3}{2}}$
$= x^{-\frac{4}{6}} \cdot x^{\frac{9}{6}}$
$= x^{\frac{5}{6}}$
$= \sqrt[6]{x^5}$

$\sqrt[6]{64^5} = \left(\sqrt[6]{64}\right)^5$
$= 2^5$
$= 32$

2.2 Simplifying Radical Expressions

1. (a) the quantity under the radical sign, e.g., $25x^2$ in $\sqrt{25x^2}$
  (b) a number expressing a power, e.g., 2 in $4x^2$, 3 in $\sqrt[3]{27}$
  (c) expressions involving roots with different indexes and/or bases

3. (a) $\sqrt{4 \cdot 5}$
  (b) $\sqrt{4} \cdot \sqrt{5}$
  (c) $\sqrt{4 \cdot 5} = \sqrt{4} \cdot \sqrt{5}$

5. Like radicals have the same root index and the same radicand. $\sqrt[4]{3x^3}$ and $5\sqrt[4]{3x^3}$ are like radicals; $\sqrt{5x^3}$ and $5x\sqrt[3]{3x^3}$ are not.

7. no; no

9. (a) $4\sqrt{15}$ (b) $2\sqrt[4]{2}$ (c) $3\sqrt[3]{10}$

   (d) $-3\sqrt[3]{3}$ (e) $4\sqrt[5]{3}$ (f) $\dfrac{3}{10}\sqrt{2}$

   (g) $\dfrac{-2}{3}\sqrt[3]{2}$

11. (a) 11 (b) $7x$ (c) $2x$
   (d) 2 (e) 2 (f) $3x$

13. 24 yd^2

2.3 Multiplying and Dividing Radical Expressions

1. (a) $5, \sqrt{7}$ (b) distributive (c) $\sqrt{3}$

3. (a) $5\cdot 6\sqrt[3]{6}$ simplifies to $30\sqrt[3]{6}$

   (b) $\dfrac{30\sqrt[3]{15}}{5}$ simplifies to $6\sqrt[3]{15}$

5. This expression contains a radical, $\sqrt[3]{5}$, in the denominator.

7. *example:* To rationalize the denominator, you want to make the denominator equal to $\sqrt[4]{3^4}$, or 3. If you multiply the fraction by $\dfrac{\sqrt[4]{3}}{\sqrt[4]{3}}$, the denominator becomes $\sqrt[4]{3^2}$. To get 3^4 under the radical sign, you need to multiply the fraction by $\dfrac{\sqrt[4]{3}\cdot\sqrt[4]{3}\cdot\sqrt[4]{3}}{\sqrt[4]{3}\cdot\sqrt[4]{3}\cdot\sqrt[4]{3}}$ or $\dfrac{\sqrt[4]{3^3}}{\sqrt[4]{3^3}}$.

9. (a) $5\sqrt{8}\cdot 7\sqrt{6} = 5(7)\sqrt{8}\sqrt{6}$
   $= 35\sqrt{48}$
   $= 35\sqrt{2\cdot 2\cdot 2\cdot 2\cdot 3}$
   $= 35(4)\sqrt{3}$
   $= 140\sqrt{3}$

   (b) $\dfrac{9}{\sqrt[3]{4a^2}} = \dfrac{9\cdot\sqrt[3]{2a}}{\sqrt[3]{4a^2}\cdot\sqrt[3]{2a}}$
   $= \dfrac{9\sqrt[3]{2a}}{\sqrt[3]{8a^3}}$
   $= \dfrac{9\sqrt[3]{2a}}{2a}$

11. *example:* When you apply the FOIL rule to $(\sqrt{m}+3)\,(\sqrt{m}-3)$, you get

$m - 3\sqrt{m} + 3\sqrt{m} - 9$ or $m-9$. The product of the outside terms cancels out the product of the inside terms, and multiplying a radical first term by itself eliminates the radical sign.

13. (a) $\dfrac{\sqrt{30}}{6}$

   (b) $\dfrac{(8-\sqrt{10})(\sqrt{7}+3\sqrt{3})}{-20}$ or

   $\dfrac{8\sqrt{7}+24\sqrt{3}-\sqrt{70}-3\sqrt{30}}{-20}$

   (c) $\dfrac{5\sqrt[3]{12n}\left(\sqrt[3]{100n^{10}}\right)}{140n^5}$ or $\dfrac{\sqrt[3]{150n^2}}{14n^2}$

   (d) $\dfrac{\sqrt[4]{x^2}}{x^2}$

3.1 Simplifying Algebraic Expressions

3. (a) associative:
   $(11-7)-14 = 4-14$
   $\qquad\qquad\qquad = -10$
   $11-(7-14) = 11-(-7)$
   $\qquad\qquad\qquad = 18$
   Therefore, $(11-7)-14 \neq 11-(7-14)$
   (b) commutative:
   $15-7 = 8 \qquad\qquad 7-15 = -8$
   Therefore, $15-7 \neq 7-15$

5. *example:* $2\cdot b$ is the same as $2b$, so it is really only one term. The distributive property only applies when an expression in parentheses contains two or more terms that are connected by addition or subtraction.

7. (a) like (b) unlike (c) like
   (d) like (e) unlike

9. (a) The negative sign means that the amount in parentheses is to be multiplied by -1. Another way to write the expression would be $-1(2x-4)$.
   (b) $-2x+4$ or $4-2x$

11. (a) $6a$ (b) $11ab$ (c) $10a-6$
   (d) $15a^2-12a$ (e) 12

13. (a) $15(30+x)$ ft^2
   (b) $(15\cdot 30)+(15\cdot x)$ ft^2 = $450+15x$ ft^2
   (c) $15(30+x) = 450+15x$
   (d) distributive property

3.2 Solving Linear Equations and Formulas

1. **(a)** If a, b, and c are real numbers and $a = b$, then $a + c = b + c$ and $a - c = b - c$.

   (b) If a, b, and c are real numbers and $a = b$, then $ca = cb$ and $\dfrac{a}{c} = \dfrac{b}{c}$, $c \neq 0$.

3. **(a)** -6 is a solution to the equation.

   (b) -6 is not a solution to the equation.

5. **(a)** $x = x$, identity **(b)** $x = -9$

   (c) $y = 12$ **(d)** $k = 0.06$

   (e) $p = -8$ **(f)** $s = -20$

   (g) $s = \dfrac{8}{3}$ **(h)** $8 = 12$, null set

   (i) $k = \dfrac{15}{8}$ **(j)** $a = \dfrac{15}{4}$ **(k)** $a = \dfrac{21}{5}$

7. **(a)** $m = \dfrac{E}{c^2}$

   (b) $w = T - ma$

   (c) $C = \dfrac{5}{9}(F - 32)$

   (d) $I = \dfrac{2K - Mv^2}{w^2}$

9. 621

3.3 Applications of Equations

1. **(a)** complementary

   (b) isosceles

   (c) principle

   (d) average

   (e) markdown

3. **(a)**

Type of Question	Number	Value of Each Question	Total Value
Multiple choice	x	5	5x
True/false	3x	2	6x
Essay	$x - 2$	10	10x – 20
Fill-in	x	5	5x

   (b) true/false

   (c) $26x - 20$

5. **(a)**
$$0.09x + 0.08(2{,}000 - x) = 400$$
$$100(0.09x + 0.08(2{,}000 - x)) = 100(400)$$
$$9x + 8(2{,}000 - x) = 40{,}000$$
$$9x + 16{,}000 - 8x = 40{,}000$$
$$x = 24{,}000$$

   (b)
$$0.2(5) + 0.6x = 0.4(5 + x)$$
$$10(0.2(5) + 0.6x) = 10(0.4(5 + x))$$
$$2(5) + 6x = 4(5 + x)$$
$$10 + 6x = 20 + 4x$$
$$2x = 10$$
$$x = 5$$

7. 30 mi

9. Jakob is 16 years old and David is 18 years old.

11. 3 min from the time the officer leaves

13. 2 ft

3.4 The Rectangular Coordinate System

3.

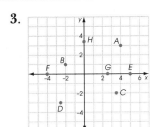

5. **(a)** 1990–1991, 1994–1995

   (b) 1990–1991

   (c) 1993

   (d) approximately 1,100,000 barrels

7. **(a)** $(2, -1)$ **(b)** no **(c)** yes

9.

4.1 Multiplying Polynomials

3. All have 6 partial products.

   (a) $x^3 + 7x^2 + 9x - 9$

   (b) $y^3 + 2y^2 - 23y - 40$

   (c) $p^3 - 3p^2 - 11p - 2$

   (d) $x^3 + \dfrac{5}{4}x^2 - \dfrac{1}{16}$

   (e) $b^4 - 2b^2 - 16a^2c^2 + 8ac$

   (f) $a^2 - b^2 + ac - bc$

5. All have 9 partial products.

   (a) $x^4 + 2x^3 + 4x^2 + 3x + 2$

   (b) $x^4 + 2x^3 + 3x^2 + 2x + 1$

   (c) $4x^4 + 8x^3 + 12x^2 + 8x + 4$

7. **(a)** $8x^2 + 24x + 16$ **(b)** $-2a^2 - 2a + 60$

   (c) $4x^2 + 32x + 48$ **(d)** $-2m^2 + 9m - 9$

   (e) $8x^2 + 2y^2 - 8xy$ **(f)** $27x^2 + 18x + 3$

   (g) $4x^3 + 28x^2 + 36x - 36$

   (h) $-2a^2 + 2b^2 - 2ac + 2bc$

   (i) $4y^3 + 8y^2 - 92y - 160$

   (j) $-x^4 + 2x^3 - x$

9. $A = P(r^3 + 3r^2 + 3r + 1)$

11. $A = 2x^2 - 4x + 2$

4.2 The Greatest Common Factor and Factoring by Grouping

1. prime-factored
3. (a) $6 - 4t$ can be further factored to $2(3 - 2t)$.
   (b) $3t - 2t^2$ can be further factored to $t(3 - 2t)$.
5. (a) $2 \cdot 7^2$
   (b) $13 \cdot 5^2$
   (c) $2^5 \cdot 3^2$
7. (a) $2(x + 4)$ (b) $2x(x - 3)$
   (c) $3y^2(y + 1)$ (d) not possible
   (e) $13ab^2c(c^2 - 2a^2)$ (f) not possible
   (g) $5t^2(5t^4 - 2t + 1)$
   (h) $9x^7y^3(5x^3 - 7y^4 + 9x^3y^7)$
   (i) $u^4v^3(48u^2v^3 - 16v - 3u^2)$
9. (a) $(x + y)(4 + t)$ (b) $(a - b)(r - s)$
   (c) $(m + n)(m + n + 1)$ (d) $-x(a - b)(b + c)$
11. (a) $2b(2 - d)(c + d)$
    (b) $x(x - y)(x - 2y + z)$

4.3 Factoring Trinomials and Difference of Squares

1. (a) a whole number that exactly divides another whole number. If polynomials are multiplied together, then each polynomial is a factor of the product.
   (b) an algebraic expression consisting of two or more terms
   (c) a polynomial consisting of three terms
   (d) a polynomial consisting of two terms
   (e) a part of a polynomial that is separated from the rest by a plus or minus sign. In $x + 10$, x and 10 are both terms.
   (f) a numerical multiplier of the variables in an algebraic term. In $3x - 2y$, 3 and 2 are coefficients.
   (g) a trinomial that is the product of two equal factors; e.g., $x^2 - 2x - 1$ is a perfect square trinomial because it equals $(x - 1)(x - 1)$.
   (h) in algebra, an expression of the form $a^2 - b^2$. It can be factored to be $(a - b)(a + b)$.
3. (a) $(2m + 5)(2m - 3)$ (b) $(3n - 5)^2$
   (c) $(5v - 4)^2$ (d) $(6a + 11)(6a - 11)$
   (e) $(3c + 7d)(3c - 7d)$ (f) $(2h + 5)(h - 4)$
   (g) $(2k - 5)(3k - 4)$ (h) $2(x^2 + 2)(x^2 + 5)$
5. 11, 29, 16
7. $3x + 5$ and $x + 11$
9. (a) difference of squares; $(3p + 2)(3p - 2)$
   (b) neither; expression cannot be factored

(c) perfect square; $(5r + 1)^2$
(d) neither; the expression cannot be factored

13. $\left(2x + \dfrac{1}{2}\right)^2$

4.4 Sum and Difference of Two Cubes

1. 1, 8, 27, 64, 125, 216, 343, 512, 729, 1,000
3. *examples:*
   Similarities: involves finding roots; factored result has two terms
   Differences: can only factor a difference of squares, but can factor a sum and a difference of cubes; factored result for difference of squares is two binomials, but for difference of cubes it is a binomial and a trinomial
5. (a) $(x + y)(x^2 - xy + y^2)$
   (b) $(x - 2)(x^2 + 2x + 4)$
   (c) $(3x + 4y)(9x^2 - 12xy + 16y^2)$
   (d) $(3a - b)(9a^2 + 3ab + b^2)$
   (e) $(x + 3)(x^2 - 3x + 9)$
   (f) $(t - 4)(t^2 + 4t + 16)$
   (g) $(4x + 5y)(16x^2 - 20xy + 25y^2)$
7. (a) $2(2 + x)(4 - 2x + x^2)(2 - x)(4 + 2x + x^2)$
   (b) $(a - b)(a^2 + ab + b^2)(a + b)(a^2 - ab + b^2)$
   (c) $(2x^2 - 5y)(4x^4 + 10x^2y + 25y^2)$
   (d) $(5xy^2 - 6z^3)(25x^2y^4 + 30xy^2z^3 + 36z^6)$
   (e) $(3x - y)(9x^2 + 3xy + y^2)(3x + y)(9x^2 - 3xy + y^2)$
   (f) $(4x^2 + 5y)(16x^4 - 20x^2y + 25y^2)$
   (g) $(a - 3b)(a^2 + 3ab + 9b^2)(a + 3b)(a^2 - 3ab + 9b^2)$
   (h) $(2x - 5y^2)(4x^2 + 10xy^2 + 25y^4)$
   (i) $6a(2x^2 - y)(4x^4 + 2x^2y + y^2)$
   (j) $3(a - 3b)(a^2 + 3ab + 9b^2)(a + 3b)(a^2 - 3ab + 9b^2)$
   (k) $(x - a^2y)(x^2 + a^2xy + a^4y^2)$
9. (a) $2x^3(4 - x)(16 + 4x + x^2)$
   (b) $(3x - 1)(9x^2 + 3x + 1)$
   (c) $5(a - 3b)(a^2 + 3ab + 9b^2)$
   (d) $(x - y)(a + z)(a^2 - az + z^2)$
   (e) $(xy^2 + 5ab^2)(x^2y^4 - 5ab^2xy^2 + 25a^2b^4)$
   (f) $(4x + 1)(16x^2 - 4x + 1)$
11. (a) $V = \dfrac{4}{3}\pi r_1^3 - \dfrac{4}{3}\pi r_2^3$
    (b) $\dfrac{4}{3}\pi(r_1 - r_2)(r_1^2 + r_1r_2 + r_2^2)$
    (c) 165.46 in.^{3}

4.5 Dividing Polynomials by Binomials

1. dividend is $x^2 + 10$; divisor is $x - 4$; quotient is $x + 4$; remainder is 26
3. (a) $(3p^3 - 2p^2 + 6p - 8) \div (p - 2)$
   (b) statement is correct
5. $b^2 + 6b - 18 = (b + 8)(b - 2) - 2$
7. $3q^3 + 5q^2 + 4q - 3 = (3q - 1)(q^2 + 2q + 2) - 1$
9. $12x^3 + 0x^2 + 1x + 1 = (2x + 1)(6x^2 - 3x + 2) - 1$
11. $6x^2 - 24 = (2x + 4)(3x - 6)$;
    length $= (3x - 6)$ cm
13. $336r^3 + 1{,}008r^2 + 1{,}008r + 336$
15. 700
17. yes

5.1 Rational Expressions: Finding Equivalent Forms

1. Similar: still eliminate common factors from the numerator and denominator
   Different: the expression has variable terms
3. (a) $\dfrac{(x+3)(x+1)}{(x-3)(x+2)}$ (b) $\dfrac{-4(y+1)}{3(y+3)}$ (c) $\dfrac{(3a-1)}{2}$
   (d) $-9 - r^2$ (e) -1
5. (a) Solution 4
   (b) Solution 1: denominator in line 2 should be $(2x - 3)(3x - 1)$
      Solution 2: line 4 is incorrect, since $\dfrac{2x+3}{2x-3}$ cannot be simplified by adding unlike terms
      Solution 3: line 4 is incorrect, since $\dfrac{2x+3}{2x-3}$ cannot be simplified by factoring out $2x$

5.2 Non-permissible Values

1. Because the value of the expression would be undefined
3. (a) yes
   (b) -2: yes; 2: no
   (c) yes
   (d) -4: yes; 8: no
5. $k = 4$
7. (a) $\dfrac{x-4}{x-3}$, $x = 3, -3$ (b) $\dfrac{2}{t+5}$, $t = 3, -5$
   (c) $\dfrac{x-5}{x-2}$, $x = 2, -5$ (d) $\dfrac{4b-1}{b+9}$, $b = 9, -9$
   (e) $\dfrac{11-x}{x+6}$, $x = -6, -1$
   (f) $\dfrac{x-y}{2x-y}$, $x = \dfrac{y}{2}$, $y = 2x$, $x = -\dfrac{5}{3}y$, $y = -\dfrac{3}{5}x$

9. (a) example: as velocity decreases, time increases
   (b) example: if velocity is 0, then time is undefined
   (c) example: In general, velocity is not given a negative value. In physics, velocity is assigned a negative value if the object is traveling away from a destination.

5.3 Multiplying and Dividing Rational Expressions

1. (a) A common factor is a quantity that divides evenly into two numbers.
   (b) When you multiply a number by its reciprocal, the product is 1.
   (c) The divisor is the amount you divide by.
   (d) The dividend is the amount you divide into parts.
3. (a) $\dfrac{b^2}{3a^2}$, $a \neq 0$, $b \neq 0$
   (b) $\dfrac{(c+5)(2c+3)}{c}$, $x \neq 0$, $\dfrac{3}{2}$, 7
   (c) $-\dfrac{x+2}{x-6}$, $x \neq 6$, 7, $-\dfrac{1}{2}$, 3
   (d) $\dfrac{4m-1}{2m-1}$, $m \neq -\dfrac{2}{3}$, $\dfrac{1}{2}$, $\dfrac{1}{5}$, -4
5. (a) $\dfrac{bx}{5c^3}$, $a, b, c, x, y \neq 0$
   (b) $y + 3$, $y \neq 0, 4, -3, 7, -1$
   (c) x, $x \neq 3a, -3a, a, -a$
   (d) 1, $x \neq -\dfrac{4}{3}$, $\dfrac{1}{2}$, $-\dfrac{3}{2}$, 4, $\dfrac{2}{3}$, $-\dfrac{1}{2}$
   (e) $\dfrac{3n-2}{n+4}$, $n \neq -\dfrac{1}{5}$, 5, $-\dfrac{1}{2}$, $-\dfrac{3}{2}$, -4, $\dfrac{2}{3}$, -3,
7. Caitlin forgot to list the non-permissible values.
   $s \neq -\dfrac{5}{2}$, 4, $\dfrac{3}{8}$, $\dfrac{2}{3}$, $\dfrac{8}{3}$
   She could also have factored out opposites to simplify the expression further.
   $\dfrac{(3s-2)(-1)(8s-3)}{(-1)(3s-2)(3s-8)} = \dfrac{8s-3}{3s-8}$
9. (a) $-\dfrac{x+3}{2x-5}$, $x \neq -\dfrac{4}{3}$, -3, $\dfrac{5}{2}$, $-\dfrac{7}{3}$
   (b) 5
   (c) This would make the width negative.

5.4 Adding and Subtracting Rational Expressions

1. (a) the simplest expression that has both denominators as factors

(b) an algebraic expression that represents the ratio of two expressions

(c) a value for a variable that makes the denominator equal to 0

3. (a) 40 **(b)** a^2 **(c)** $6y$

(d) $x^2 - 1$ **(e)** $15(x+3)$

5. (a) $\dfrac{5a-3}{(a+6)(a-5)}$ **(b)** $\dfrac{a-48}{(a+7)(a-4)}$

(c) $\dfrac{3t-4}{6(t+3)}$ **(d)** $\dfrac{2q^2-12q-1}{(9+q)(9-q)}$

(e) $\dfrac{8y}{(y-4)(y+4)}$ **(f)** $\dfrac{9a^2+16a+1}{(6a+1)(a+2)(a+3)}$

(g) $\dfrac{3(d-14)}{d-6}$ **(h)** $\dfrac{x^2+4x-2}{4(x+3)(x-2)}$

(i) $\dfrac{k^2}{k-1}$ **(j)** $\dfrac{2x+9}{x+3}$

(k) $\dfrac{-x^2+5x-2}{(x-1)(2x-1)}$

7. $\dfrac{x^3+17x^2+7x-4}{x(x+2)(3x-1)}$

9. Tabor could have simplified both of the original expressions before he added. His answer is correct, but it is not in simplest form.

$$\frac{5x+25}{(x+5)(x-2)} = \frac{5(x+5)}{(x+5)(x-2)} = \frac{5}{x-2}$$

$$\frac{2(x-4)}{(x-4)(x+3)} = \frac{2}{x+3}$$

$$\frac{5}{x-2} - \frac{2}{x+3} = \frac{3x+19}{(x-2)(x+3)}$$

5.5 Solving Rational Equations

1. When you multiply the numerator by a multiple of the denominator, the numerator itself becomes a multiple of the denominator. When you eliminate common factors from the expression, the denominator becomes 1.

5. 6: **(a)** none   **(b)** $x \neq 0$

     (c) $x \neq 0$ **(d)** $x \neq 0$

     7: **(a)** $x \neq -1$ **(b)** $y \neq 3$

     (c) $x \neq \dfrac{3}{2}$ **(d)** $n \neq 7, 25$

     8: **(a)** $x \neq \pm 1$ **(b)** $y \neq \pm 3$

     (c) $t \neq 0, 5$ **(d)** $x \neq 0, \pm 1$

7. (a) $x = 5$ **(b)** $y = 13$

(c) $x = 6$ **(d)** $n = -20$

9. 20,000 rolls

11. Jamilla forgot to multiply the -2 on the left side of the equation by the LCD.

$$-2 \cdot 3(x+1) = -6(x+1)$$
$$= -6x - 6$$

Now the simplified equation becomes:

$$9 - 6x - 6 = 5$$
$$3 - 6x = 5$$
$$-6x = 2$$
$$x = -\frac{1}{3}$$

13. $\dfrac{5x+10}{x^2-4} \cdot 2 = \dfrac{15}{12}$, so $x = 10$

6.1 Differences Between Relations and Functions

3. (a) $(0, 5)$ $(1, 4)$ $(2, 3)$ $(3, 2)$ $(4, 1)$

(b) yes, since this is a one-to-one mapping

5. *example:* same: all are binary relations; different:

(a) is one-to-one

(b) is many-to-one

(c) is one-to-many

7. (a)   **(b)**

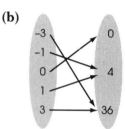

9. (a) $(3, 1)$ **(b)** $(3, 4)$

(c) $(2, 2)$ **(d)** $(3, 4)$ and $(4, 3)$

11. (a) A $(-7, 2)$ B $(7, 2)$

(b) many-to-one; *example:* there are 2 possible values for x when $y = 2$

6.2 Describing Functions

1. (a) distance traveled depends on time

(b) volume depends on pressure

(c) area depends on length

3. (a) $\left(-1, -\dfrac{17}{3}\right) \left(\dfrac{45}{2}, 10\right)$ **(b)** $(-4, 44)$ $(3, 23)$

(c) $(4, 7)$ $(12, 11)$ **(d)** $(2, 8)$ $(4, 80)$

(e) $(-2, -15)$ $(1, 0)$ **(f)** $(1, -4)$ $(-3, -6)$

5. 9

7. (a) $y = \sqrt{x} - 2$ **(b)** $y = 3^x$

(c) \$2 per hour or any portion thereof

9. 15

6.3 Function Notation

1. *example:* to help understand the relationship between an independent and dependent variable; a consistent way to represent that relationship

3. (a) $f(x) = \sqrt{x} - 3$ (b) $f(x) = \dfrac{x^3}{x-1}$

5.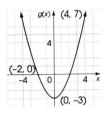

7. (a) -3 (b) $2, -8$ (c) 4

9. $d(1) = 12.5$; $d(7) = 612.5$; $d(12) = 1{,}800$

11. $4, -4$

13. (a) -6 (b) 6 (c) -11
     (d) $2h$ (e) -5 (f) -1
     (g) -1 (h) 47 (i) $2a + h$
     (j) 1

6.4 Domain and Range of Relations

1. the set of first coordinates of the ordered pairs

3. the set of second coordinates of the ordered pairs

5. The graph is a dotted line without bounds. There is an open point at $x = 5$, since there is no corresponding y-value. The other points are located at all possible integer values for x and are closed points.

7. The graph is a solid, horizontal line extending without bounds. The y-intercept is at $(0, 7)$.

9. (a) $\{x \mid x \in R\}$; $\{y \mid y \in R\}$
     (b) $\{x \mid x \in R\}$; $\{y \mid y \le 3, y \in R\}$
     (c) $\{x \mid 0 \le x \le 4, x \in R\}$; $\{y \mid 1 \le y \le 5, y \in R\}$
     (d) $\{x \mid x \ne 0, x \in R\}$; $\{y \mid y > 0, y \in R\}$

11.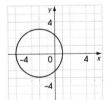

7.1 Linear Functions

1. When a graph crosses the x-axis, the value of y is zero. By substituting zero for y and using the value for the slope and y-intercept, you can solve for x.

3. (a) domain: x is a real number ≥ 0; range: y is a real number ≥ 1
     (b) domain: x is a real number; range: $y = 2$
     (c) domain: $x = -4, -2, 0, 2$; range: $y = 0, -1, -2, -3$

5. (a) $y = \dfrac{1}{2}x - 2$ (b) $y = \dfrac{2}{3}x + 2$
     (c) $y = -\dfrac{5}{4}x + 5$ (d) $y = -x$

7. (a)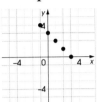

     (b)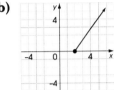

     (c)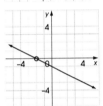

9. (a) domain: $x = 0, 1, 2, 3, 4$
        range: $y = 6, 2, -2, -6, -10$
     (b) domain: $x = 1, 2, 3, 4, 5$
        range: $y = 1.5, 2, 2.5, 3, 3.5$
     (c) domain: x is a real number
        range: y is a real number
     (d) domain: x is a real number ≥ 0
        range: y is a real number ≥ 1

11. (a) -1 (b) 0 (c) 1.5

13. (a) yes
     (b) no; if x and y are positive, then $\dfrac{y-0}{0-x} = \dfrac{positive}{negative}$, which must result in a negative slope

15. (a) and (d)

7.2 Direct and Partial Variation

1. (a) where y varies directly with x, and the y-intercept is at the origin: $y = kx$
     (b) where y varies partially with x, but the y-intercept is not at the origin: $y = kx + b$
     (c) the constant k in the equation for direct variation $y = kx$; the slope
     (d) the output or y-values of a relation
     (e) the input or x-values of a relation

(f) the unchanging cost for a given situation
(g) cost which depends on another variable
3. (a) $k = 2$ **(b)** $k = 12$ **(c)** $k = 5$
5. (a) $y = 50$ **(b)** $y = 3.6$
7. (a) 12 **(b)** 32 **(c)** 17 **(d)** 59
9. (a) $M = kV$ **(b)** 480 g
11. \$6,625
13. 0.96 m or 96 cm
15. *example:* Since the fixed cost is in dollars, the agent should have used 0.10 for the variable cost, not 10. Actual cost is $\$40 + 3{,}260\cent$ or $\$40 + \$32.60$ or \$72.60.

7.3 Rate of Change and Slope of a Line

1. *example:* The slope is the distance a line rises for each unit of horizontal distance it travels. You calculate it by dividing the vertical distance a line travels between two points by the horizontal distance. Use the formula $\frac{rise}{run}$ if you can count squares, or the formula $\frac{y_1 - y_2}{x_1 - x_2}$ if you can't. The slope of the line $y = x$ is 1. If a line rises more steeply than $y = x$, the slope is more than 1. If the line rises less steeply than $y = x$, the slope is between 0 and 1. If the line is horizontal, the slope is 0. If the line falls to the right, the slope is negative.

3. (a) $\frac{3}{5}$ **(b)** $-\frac{5}{3}$ **(c)** $-\frac{1}{6}$
(d) $-\frac{4}{8} = -\frac{1}{2}$ **(e)** $\frac{8}{4} = 2$ **(f)** $\frac{4}{2} = 2$
(g) $\frac{3}{6} = \frac{1}{2}$ **(h)** $\frac{4}{2} = 2$

5. *example:* Horizontal lines don't rise, so the rise is always $\frac{0}{run}$. Vertical lines have no run, so the slope is $\frac{rise}{0}$, and division by 0 is undefined.

7. (a) $y = \frac{2}{3}x + \frac{11}{3}$ **(b)** $y = -\frac{4}{3}x + \frac{17}{3}$
(c) $y = -\frac{3}{2}x + \frac{13}{2}$ **(d)** $y = -4x - 9$
9. *example:* Expressing the slope as an improper fraction shows the relationship between rise (in the numerator) and run (in the denominator).
11. *example:* points are collinear because slope AB = slope $BC = \frac{2}{3}$; points are collinear because all three sets of coordinates satisfy the equation of the line, which is $2x - 3y + 14 = 0$

7.4 Determining the Equation of a Line

1. *example:* The x-coordinate of the y-intercept is always 0.
3. (a) slope = 3, y-intercept = 6
(b) slope $= -\frac{3}{2}$, y-intercept = 3
5. (a) $2x - y - 5 = 0$ **(b)** $2x + 3y - 7 = 0$
(c) $x - 2y + 5 = 0$ **(d)** $3x + y + 4 = 0$
7. (a) $y = x + 1$ **(b)** $y = 3$
(c) $x = 4$ **(d)** $y = -x - 2$
9. (a) $5x + 4y + 12 = 0$ **(b)** $x + 4y - 4 = 0$
(c) $4x - 3y - 12 = 0$
11. $\frac{32}{3}$ or $10\frac{2}{3}$
13. (b) $y = \frac{13}{2}x$ or $y = 6.5x$
(c) *example:* The y-intercept is $(0, 0)$, so Kara is paid by the hour, with no flat rate. The slope is 6.5, so Kara is paid \$6.50/h.

8.1 Solving Linear Systems by Graphing

1. (a) two or more linear equations that are considered together
(b) the solution is the intersection of the graphs of the linear equations
(c) Substitute the coordinates of the solution into each linear equation to verify it holds true.
3. (a) $(2, -10)$ **(b)** $(-5, 4)$ **(c)** $(-1, -1)$
5. (a) Line 1: slope: 1; y-intercept: -2
Line 2: slope: $\frac{8}{9}$; y-intercept: -1
solution: $(9, 7)$
(b) Line 1: slope is undefined; no y-intercept
Line 2: slope: 0; y-intercept: 7
solution: $(4, 7)$
(c) Line 1: slope: $-\frac{5}{8}$; y-intercept: 6
Line 2: slope: $-\frac{5}{8}$; y-intercept: 6
number of solutions is infinite because the system is dependent
7. (a) $(3, 4)$ **(b)** $(5, -2)$
9. (a) yes **(b)** yes
11. (a) *m:* anything except $\frac{3}{4}$; *b:* anything
(b) $m = \frac{3}{4}$; *b:* anything except -2
(c) $m = \frac{3}{4}$; $b = -2$
13. $(2, -8)$
15. any ordered pairs where $x = y$, and x and y are real numbers

8.2 Solving Linear Systems by Elimination

1. **(a)** $x = 3$, $y = 2$ **(b)** $x = 4$, $y = 1$
   (c) $x = 7$, $y = 15$
3. $a = -1$, $b = 1$
5. $A = 3$, $B = 5$
7. $(3, 6)$, $(5, -2)$, $(-5, 3)$
9. Let y represent the number of tickets sold at $4.50. Let s represent the number of tickets sold at $6.00.

$$\begin{cases} 4.5y + 6s = 792 \\ y + s = 152 \end{cases}$$

$y = 80$, $s = 72$

8.3 Solving Three Equations in Three Variables

1. *example:*

$$10a + 8b - 2c = 20$$
$$+\ 3a - 6b + 2c = 15$$
$$13a + 2b = 35$$

   Multiply the first equation by 2 to make the last term $-2c$. Add the two equations to eliminate the $2c$'s.
3. **(a)** $(2, -1, 3)$ is not a solution
   (b) $(4, 2, 0)$ is not a solution
   (c) $(-10, -9, 5)$ is not a solution
   (d) $(-2, 5, 3)$ is a solution
5. **(a)** $(2, 1, -1)$ **(b)** no solution
   (c) $(-5, -2, 4)$ **(d)** $(3, -2, -4)$
   (e) infinite number of solutions
   (f) $(2, 4, -3)$
7. 10 h in class, 30 h studying, 20 h at work
9. $15,000 at 5%, $22,000 at 7%, and $13,000 at 4%.

8.4 Solving Linear Systems by Matrices

1. **(a)** matrix
   (b) elements
   (c) 3, 4
   (d) row
   (e) augmented
   (f) main diagonal
3. **(a)** 2 rows, 3 columns
   (b) 3 rows, 4 columns
5. **(a)** $x - y = -10$; $y = 6$; solution: $x = -4$, $y = 6$
   (b) $x - 2y + z = -16$; $y + 2z = 8$; $z = 4$; solution: $x = -20$, $y = 0$, $z = 4$
7. **(a)** exchange the first and second rows;

$$\begin{bmatrix} 1 & -4 & | & 4 \\ -3 & 1 & | & -6 \end{bmatrix}$$

(b) add three times the first row to the second row; $\begin{bmatrix} 1 & -4 & | & 4 \\ 0 & -11 & | & 6 \end{bmatrix}$

9. **(a)** inconsistent
   (b) inconsistent
   (c) dependent
   (d) dependent
   (e) $x = 0$, $y = 1$, $z = 3$
   (f) $x = 8$, $y = 4$, $z = 5$
   (g) inconsistent
   (h) dependent
   (i) $x = -4$, $y = 8$, $z = 5$
   (j) $x = -1$, $y = 2$, $z = -2$
   (k) dependent
   (l) inconsistent
11. *example:* founder's circle: 100, box seats: 300, promenade: 400

8.5 Solving Linear Systems by Determinants

1. *examples:*
   (a) the entries along a horizontal line in a matrix
   (b) the entries along a vertical line in a matrix
   (c) a matrix of the coefficients with an added column containing the constants
   (d) a matrix where the number of rows equals the number of columns
   (e) the determinant of a sub-matrix obtained by striking out the row and the column in which a chosen element lies
3. Form the augmented matrix. Find D_x by substituting the constant values for the x-values. Evaluate the determinant. By Cramer's rule, $x = \dfrac{D_x}{D}$.
5. $D = -3(4 - 1) + 0 + 2(-1 - 6)$
   $= -3(3) + 2(-7)$
   $= -9 - 14$
   $= -23$
7. **(a)** $x = -1$, $y = 3$
   (b) $x = -3$, $y = -1$
   (c) $x = 3$, $y = 4$
   (d) $x = 1$, $y = 1$, $z = 2$
   (e) $x = 0$, $y = 2$, $z = 2$
   (f) inconsistent system
   (g) dependent system
   (h) $x = -2$, $y = 3$, $z = 1$
9. $8,000 in SaveTel, $7,000 in OilCo, and $5,000 in HiTech

11. a zero entry will result in a zero when multiplied by its minor

8.6 Solving Systems of Linear Inequalities

1. (a) true   (b) false
    (c) false   (d) false
3. not possible
5. Let L represent the length in yards.
Let W represent the width in yards.
$L > W$, $2L + 2W \geq 20$, $LW \geq 30$

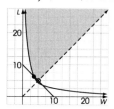

    not possible

8.7 Linear Programming

1. 55
3. (a) corner points are: $(0, 3)$ $(0, 2)$ $(4, \frac{2}{3})$ $(\frac{9}{7}, \frac{30}{7})$

maximum value is $\frac{204}{7}$ at $(\frac{9}{7}, \frac{30}{7})$

  (b) The solution set is $\{(1, 2)\ (1, 3)\ (1, 4)\ (2, 2)\ (2, 3)\ (3, 1)\ (3, 2)\}$
maximum value is 28 at $(3, 2)$
5. Let g = hours spent golfing
Let b = hours spent cycling
Let C = calories burned in a week

Constraints:	Objective Function
$g + b \leq 30$	$C = 300g + 420b$
$g + b \geq 15$	
$g \geq 2b$	
$b \geq 3$	

corner points (g, b) are: $(10, 5)$ $(12, 3)$ $(27, 3)$ $(20, 10)$
maximum number of calories burned is 10,200 at $(20, 10)$
minimum number of calories burned is 4,860 at $(12, 3)$
7. Let a = number of adult bikes
Let c = number of child's bikes
Let P = profit in dollars

Constraints:	Objective Function
$a + c \leq 48$	$P = 75a + 45c$
$210a + 90c \leq 7,350$	
$a \leq 2c$	

maximum value is $\$2,910$ at $(a, c) = (25, 23)$

9.1 Functions and Operations

1. For each x-value, find the corresponding values of $f(x)$ and $g(x)$, and add these together.
3. (a) $-3x + 2$ (b) $9x + 2$
    (c) $18x^2 + 12x$ (d) $3x - 2$
5. (a) $12x + 3$ (b) $12x + 1$
    (c) $16x^2 + 8x + 1$ (d) $4x^2 + 1$
    (e) $-x^2 + 1$ (f) $16x + 5$
    (g) $9x$ (h) $12x^2$
7. $f(x) = 3x$ $g(x) = x - 2$

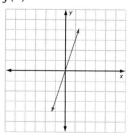 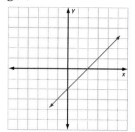

$(f - g)(x) = 2x + 2$

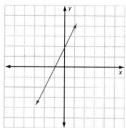

9. $16w + 104$
11. $K(t) = K(v(t))$
$= K(30 - 9.8t)$
$= 0.4(30 - 9.8t)^2$
$\approx 360 - 235t + 38t^2$
13. (a) $3\frac{3}{4}$ yards
    (b) The amount of fabric depends on the pattern size which depends on the child's chest size.

9.2 Inverse Functions

1. $f^{-1}(x) = \sqrt[3]{3(x + 1)}$
3. The inverse is not a function. This notation is reserved for functions.
5. (a) $f^{-1}(x) = \{(2, -1), (4, 3), (-7, 0), (-6, 11)\}$
    (b) $f^{-1}(x) = \dfrac{x + 2}{4}$
    (c) $f^{-1}(x) = \dfrac{1}{x - 4}$
    (d) $f^{-1}(x) = \sqrt[3]{x - 2}$

7. The remaining steps are:

$xy - 2x = 3y$
$xy - 3y = 2x$
$y(x - 3) = 2x$
$g^{-1}(x) = \dfrac{2x}{x-3}$

9. (a) $x = -12$ is a horizontal line and so passes through more than one point. Therefore, the inverse is not a function.

(b) The inverse is symmetric, and a horizontal line would pass through more than one point. Therefore, the inverse is not a function.

(c) $(3, 2)$ and $(0, 2)$ have the same y-value, so a horizontal line would pass through two points. Therefore, the inverse is not a function.

11. (a) and (d)

13. (a) $f^{-1}(x) = 2x + 6$, $f^{-1}(0) = 6$, $f^{-1}(-2) = 2$

(b) $f^{-1}(x) = \dfrac{4}{x} - 1$, $f^{-1}(0)$ is undefined,
$f^{-1}(-2) = -3$

(c) $f^{-1}(x) = \sqrt[3]{-x + 8}$, $f^{-1}(0) = 2$,
$f^{-1}(-2) = \sqrt[3]{10}$

(d) $f^{-1}(x) = \{(-2, 0), (4, -2), (0, 3)\}$,
$f^{-1}(0) = 3$, $f^{-1}(-2) = 0$

(e) $f^{-1}(x) = \dfrac{3x + 4}{2}$, $f^{-1}(0) = 2$, $f^{-1}(-2) = -1$

15. domain: $\{x \mid x \geq -1\}$,
range: $\{y \mid y \geq 0\}$ $h^{-1}(x) = x^2 - 1$, $x \geq 0$

9.3 Graphs of Quadratic Functions

1. If the graph opens up, it has a minimum value; if the graph opens down, it has a maximum value.

3. opens up; minimum of 0

5. vertex $= (1, -3)$; minimum of -3; equation of axis of symmetry is $x = 1$

7. no x-intercepts; y-intercept at $(0, -1)$

9. domain: $\{x \mid x \leq 0, x \in \mathbb{R}\}$
range: $\{y \mid y \leq 4, y \in \mathbb{R}\}$

9.4 Completing the Square

3. a represents the shape of the graph relative to the graph of $y = x$, h represents the x-coordinate of the vertex, k represents the y-coordinate of the vertex

5. (a) True

(b) False; this parabola opens upward and is wider than the basic $y = x^2$ graph.

(c) False; this parabola opens upward and is narrower than the basic $y = x^2$ graph.

(d) True

7. (a) $y = -7(x^2 + 2x) + 9$

(b) $f(x) = -2.3(x^2 - 4x) - 5.4$

(c) $y = 4\left(x^2 - \dfrac{1}{2}x\right) + 5$

(d) $y = 15\left(x^2 - \dfrac{1}{3}x\right) + 2$

(e) $y = \dfrac{1}{3}\left(x^2 + \dfrac{3}{5}x\right) - \dfrac{4}{9}$

(f) $f(x) = \dfrac{3}{8}\left(x^2 + \dfrac{2}{3}x\right) + \dfrac{1}{2}$

9. (a) $y = (x + 5)^2 + 3$

(b) $f(x) = 2(x + 2)^2 - 1$

(c) $y = 3(x - 3)^2 - 26$

(d) $y = \left(x - \dfrac{3}{2}\right)^2 + 4\dfrac{3}{4}$

11. (a) vertex $= (1, -2)$; y-intercept $= (0, 1)$

(b) vertex $= (-4, 10)$; y-intercept $= (0, 6)$

(c) vertex $= (-3, -7)$; y-intercept $= (0, -1)$

(d) vertex $= \left(-2, 3\dfrac{1}{6}\right)$; $y -$ intercept $= \left(0, \dfrac{1}{2}\right)$

9.5 Translations of Quadratic Functions

1. (a) quadratic

(b) vertex

(c) axis of symmetry

3. (a)

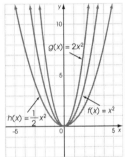

(b)

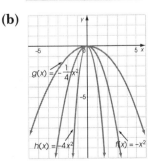

5. **(a)** to obtain $g(x)$, move $f(x)$ up 3 units; to obtain $h(x)$, move $f(x)$ down 1 unit

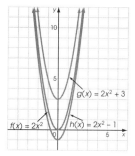

(b) to obtain $g(x)$, move $f(x)$ to the left 2 units; to obtain $h(x)$, move $f(x)$ to the right 3 units

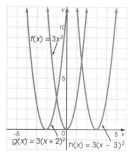

7. **(a)** $x = 3$, $(6, 18)$
   (b) $x = 0$, $(2, 19)$
   (c) $x = -2$, $(-8, 15)$
   (d) $x = 5$, $(11, 42)$
9. **(a)** vertex $(1, 2)$, axis of symmetry $x = 1$, opens upward
   (b) vertex $(2, -1)$, axis of symmetry $x = 2$, opens downward
   (c) vertex $(-3, -4)$, axis of symmetry $x = -3$, opens upward

10.1 Solving Quadratics by Graphing

1. equation/root; function/zero; graph/x-intercept
3. no, since a parabola can cross the x-axis two times (at most).
5. **(a)** $2x^2 - 6x + 0 = 0$ or $x^2 - 3x + 0 = 0$; 3 or 0
   (b) $x^2 + 5x - 14 = 0$; -7 or 2
   (c) $x^2 - 13x + 5 = 0$; (estimated) 12.6 or 0.4
   (d) $2x^2 + 5x - 11 = 0$; (estimated) 1.4 or -3.9
   (e) $3x^2 + 11x + 4 = 0$ (estimated) -0.4 or -3.3
   (f) $x^2 - 2x - 15$; 5 or -3
7. Leticia made an error when she expanded $(x + 2)^2$ and $(x + 4)^2$ in line 2 of her solution. From line 2, the solution should be:
$$x^2 + (x^2 + 4x + 4) + (x^2 + 8x + 16) = 155$$
$$3x^2 + 12x - 135 = 0$$
$$f(x) = 3x^2 + 12x - 135$$

The x-intercepts are at 5 and -9.
If x is 5, then $x + 2$ is 7 and $x + 4$ is 9.
If x is -9, then $x + 2$ is -7 and $x + 4$ is -5.
9. 23 and 24 or -23 and -24

10.2 Solving Quadratics by Factoring

1. **(a)** a polynomial equation of degree 2 of the general form $ax^2 + bx + c = 0$
   (b) a function that can be written in the form $ax^2 + bx + c$, where a, b, c are real numbers and $a \neq 0$
   (c) The graph of $f(x) = ax^2 + bx + c$ is a parabola. It opens upward for $a > 0$ and downward for $a < 0$.
   (d) substituting the value of x into the original function/equation and seeing if it is true
   (e) the zeros of a quadratic equations
   (f) the values of x for which $f(x)$ or y is zero
   (g) the x-coordinate of a point where the graphed function touches the x-axis

3. **(a)** $0, 5$ **(b)** $-2, 5$ **(c)** $7, 4$
   (d) $3, -1$ **(e)** $\dfrac{5}{3}, -\dfrac{1}{3}$ **(f)** $6, -1$

5. Multiples of these functions are also acceptable answers.
   (a) $f(x) = x^2 - 8x + 15$
   (b) $f(x) = x^2 + \dfrac{3}{2}x - 1$
   (c) $f(x) = x^2 - \dfrac{6}{5}x + \dfrac{9}{25}$
   (d) $f(x) = x^2 + \dfrac{31}{20}x + \dfrac{3}{5}$
   (e) $f(x) = x^2 + \dfrac{6}{7}x$
   (f) $f(x) = x^2 - \dfrac{25}{4}$

7. 27 and 28 or -27 and -28
9. perimeter is $4(3x - 5)$ or $12x - 20$

10.3 Complex Numbers

3. A complex conjugate has the same real number part as the original number and the opposite complex coefficient part as the

original number. It is used to remove the complex number from a denominator by resulting in an addition of squares, $(a + bi)(a - bi) = a^2 + b^2$, which is always a real number.

5. **(a)** $3i$ **(b)** $\sqrt{11}i$ **(c)** $2\sqrt{6}i$

   (d) $-6\sqrt{2}i$ **(e)** $45i$ **(f)** $\frac{5}{3}i$

   (g) -30 **(h)** $-\frac{5}{8}$

7. **(a)** $12 - 8i$ **(b)** $2 - 68i$ **(c)** $-2 - 2\sqrt{2}i$

   (d) $-18 + 128i$ **(e)** $22 + 29i$

   (f) $-12 + 28\sqrt{3}i$

9. **(a)** -1 **(b)** $-i$

   (c) i **(d)** 1

11. $a = 7, b = 1$

10.4 The Quadratic Formula

1. **Step 1**: Factor a out of the terms ax^2 and bx. **Step 2**: Determine the value needed to complete the square by dividing the coefficient of x by 2 and squaring. Add and subtract this amount so you do not change the value of the equation. **Step 3**: To complete the square, move the last term from inside the parentheses to the outside by multiplying it by a. **Step 4**: Factor the perfect square trinomial.
   Step 5: Move $-\frac{b^2}{4a}$ and $+ c$ to the right side by using inverse operations. **Step 6**: Divide both sides by a. **Step 7**: Find a common denominator for terms on the right side. **Step 8**: Subtract the terms on the right side. **Step 9**: Find the square root of both sides. **Step 10**: Subtract $\frac{b}{2a}$ from both sides. **Step 11**: Add the terms on the right side.

3. **(a)** yes **(b)** no **(c)** no **(d)** no

5. **(a)** $\dfrac{-10 \pm \sqrt{120}}{2}$, $x \approx 0.5$, $x \approx -10.5$

   (b) $\dfrac{-3 \pm \sqrt{57}}{6}$, $x \approx 0.8$, $x \approx -1.8$

7. *example*: To create a new equation with the same roots as a given equation, multiply each term in the given equation by the same amount.

9. 15.2 and 16.2 or –15.2 and –16.2

10.5 Using Discriminants and Graphs

1. In this form, you can tell whether the equation can be factored. If it can't, it is

easier to apply the quadratic formula.

3. If the discriminant is positive, there are two x-intercepts. If it is 0, the vertex of the graph is on the x-axis. If it is negative, there are no x-intercepts.

5. **(a)** no real roots **(b)** two real roots **(c)** one real root

7. **(a)** $\left\{k \mid k = 4\right\}$ **(b)** $\left\{k \mid -4\sqrt{14} < k < 4\sqrt{14}\right\}$

   (c) $\left\{k \mid k < -\frac{25}{48}\right\}$ **(d)** $\left\{k \mid k < 3\right\}$

   (e) $\left\{k \mid k = \pm 4\right\}$ **(f)** $\left\{k \mid k < -\frac{25}{8}\right\}$

9. **(a)** $n = \pm 14$ **(b)** $n = \pm 9$

11. Yes, the rocket can reach a height of 120 m.

10.6 Graphs of Polynomial Functions

1. *examples*:
   (a) the coefficient of the highest-degree term
   (b) the number of times each zero occurs
   (c) a value that makes the polynomial zero
   (d) the highest exponent in the polynomial
   (e) a function having a degree of 4
   (f) a function having a degree of 5

3. Functions of even degree have arms pointing in the same direction, whereas odd degrees produce graphs pointing in opposite directions. The number of changes in direction is at most one less than the degree.

5. **(a)** (i)
   (b) (vi)
   (c) (v)
   (d) (iv)
   (e) (iii)
   (f) (ii)

7. There are 3 different zeros. The one at 0 has a multiplicity of 2 and the other two each have a multiplicity of 1.

10.7 Graphs of Rational Functions

3. (a), (d), and (e)

5. **(a)** $y = 0$ **(b)** $y = 3$ **(c)** $y = 1$
   (d) $y = 0$ **(e)** $y = 0$ **(f)** $y = 0$

7. **(a)** domain: $\{x \mid x \neq 2, x \in \mathrm{R}\}$, range: $\{y \mid y \neq 1, y \in \mathrm{R}\}$
   (b) domain: $\{x \mid x \neq 0, x \in \mathrm{R}\}$, range: $\{y \mid y \neq 1, y \in \mathrm{R}\}$

(c) domain: $\{x \mid x \neq -2, 2, x \in R\}$, range: $\{y \mid y \neq 0, y \in R\}$, symmetric with respect to y-axis

(d) domain: $\{x \mid x \neq -1, 4, x \in R\}$, range: $\{y \mid y \neq 0, y \in R\}$

9.

vertical asymptote	$x = 2, -3$
horizontal asymptote	$y = 0$
domain	$\{x \mid x \neq 2, x \neq 3, x \in R\}$
range	$\{y \mid y \neq 0, y \in R\}$
x-intercept	none
y-intercept	$-\dfrac{1}{2}$

11.1 Absolute Value Equations

1. *example:* Absolute value means the magnitude of a number, regardless of sign. For example, 6 and –6 both have an absolute value of 6. It is necessary to consider two cases because the expression inside the absolute value signs could be positive or negative.

3. (a) 8
   (b) 14

5. (a) 8   (b) 2   (c) 9   (d) –10

7. (a) 3, –3   (b) 7, –11   (c) 2, –5
   (d) 13, –3   (e) 3, –5   (f) 5, –2

9. (a) $x = -3$ (b) $x = 1, x = 5$
   (c) $x = -2$ (d) no solution

11.2 Radical Equations

1. *example:* To isolate a radical means to locate the term containing the variable under the radical sign on one side of the equal sign and have the remaining terms on the other side of the equal sign.

3. Roots must be checked to determine if they are extraneous.

5. (a) $\sqrt{4-x} - \sqrt{x+6} = 2$
       $\sqrt{4-x} = 2 + \sqrt{x+6}$

   Isolate a single radical.

   (b) $\sqrt{\sqrt{x^2-5}} = 4$ Square both sides.
       $\sqrt{x^2-5} = 16$

   (c) $\sqrt{x+2} - 3 = 9$ Isolate the radical.
       $\sqrt{x+2} = 12$

   (d) $5x = 1 + \sqrt{3-2x}$ Isolate the radical.
       $5x - 1 = \sqrt{3-2x}$

(e) $\dfrac{\sqrt{x+2}}{3} = 4x$
    $\sqrt{x+2} = 12x$

Clear the denominator by multiplying by the LCD.

7. (a) –5   (b) –17, 17
   (c) 142   (d) $\dfrac{3}{5}$
   (e) 2

9. (a) The resistance is lower on a wet road.
   (b) Accidents 3, 4, 5, 7, 8, and 9 could have occurred in wet conditions because the coefficient of friction is low. Accidents 1, 2, and 6 could have occurred in dry conditions because the coefficient of friction is higher.

Accident number	Speed of the vehicle (mph)	Length of the skid marks (ft)	Coefficient of friction
1	55	180	0.56
2	61	144	0.87
3	50	172	0.48
4	33	107	0.34
5	64	569	0.24
6	62	215	0.59
7	40	192	0.28
8	32	40	0.85
9	75	815	0.23

(c) 67 mph, 47 mph
(d) No. The speed was only $\sqrt{2}$ times as fast. This can be seen by substituting numbers into the formula. Also, the two sets of skid marks could have been made under different weather conditions.
(e) On dry concrete, the skid marks would be 172 ft, and on wet concrete, the skid marks would be 335 ft.

11.3 Rational Equations

3. (a) 0, 1, and 4
   (b) –2, 0, and 2
   (c) –1 and $\dfrac{4}{3}$
   (d) –4, –3, and 3
   (e) $-\dfrac{3}{2}, -\dfrac{1}{3}$, and 1

5. (a) Solution set $\{\dfrac{1}{3}, 2\}$
   (b) Solution set = $\{1\}$
   (c) Solution set = $\{1\}$
   (d) Solution set $\{-4, \dfrac{3}{2}\}$

(e) Solution set $= \{-\frac{1}{2}, 6\}$

7. (a) In the second line, the second term, –2, was not multiplied by the LCD.
   (b) In the second line, the first term's denominator does not cancel with the LCD. The term must be changed to $\frac{-1}{x-1}$.
   (c) In the third line, $-6x + 18$ should be $-6x - 18$.

11.4 Rational Inequalities 1

1. (a) quadratic
   (b) 2
   (c) interval

3. (a)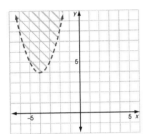

   (b)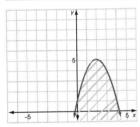

   (c)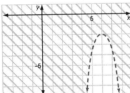

   (d)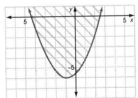

   (e)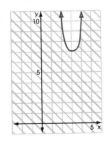

5. (a) $(-\infty, -3) \cup (5, \infty)$
   (b) $(-\infty, -4] \cup [7, \infty)$
   (c) $[-6, -4]$
   (d) $(-\infty, -8) \cup (11, \infty)$
   (e) $\left(-\frac{1}{2}, 10\right)$
   (f) $(-\infty, -10] \cup [4, \infty)$
   (g) $\left[-5, \frac{4}{3}\right]$

7. (a) $(-\infty, 0) \cup (3, \infty)$
   (b) $[-7, -1)$
   (c) $(-\infty, -8) \cup (-5, 2)$
   (d) $(-12, -3) \cup [1, \infty)$
   (e) $(-8, 2] \cup (3, 4]$
   (f) $(-\infty, -6) \cup (-6, 0) \cup (2, 10)$
   (g) $(0, 2) \cup (8, \infty)$
   (h) $(-\infty, -9) \cup [-3, -1) \cup [3, \infty)$

9. $10 \le x \le 50$ bracelets per week

11.5 Rational Inequalities 2

5. (a) $(-3, 3)$
   (b) $(-\infty, 1] \cup [11, \infty)$
   (c) $[-14, -2]$
   (d) $(-\infty, -8) \cup (14, \infty)$
   (e) R
   (f) $\left(-\infty, -\frac{3}{2}\right] \cup \left[\frac{3}{2}, \infty\right)$
   (g) $\varnothing$
   (h) $[1, \frac{11}{3}]$

7. (a) (ii)
   (b) (iii)
   (c) (i)

9. examples:
   (a) $|x + 1| \ge 4$ (b) $|x - 1| \le 5$
   (c) $|x| < 4$ (d) $|x - 2| > 3$

13. Solution set $= \{-40, 120\}$

11.6 Complex Fractions

1. example: A complex fraction has a fraction in the numerator and/or the denominator, or contains a power with a negative exponent.

   examples: $\dfrac{\frac{3}{5}}{\frac{2}{3}}, \dfrac{\frac{3ab}{c}}{4a}, \dfrac{5b^{-1}}{\frac{3b}{4}}$

3. (a) $\frac{71}{57}$ (b) $\frac{6}{7}$

5. Step 3: Factor the numerator and denominator and rewrite the fraction in

simplest terms.

$$\frac{(3y-1)(y+1)}{2(y^2-1)} = \frac{(3y-1)(y+1)}{2(y+1)(y-1)} = \frac{(3y-1)}{2(y-1)}$$

12.1 Exponential Equations

3. **(a)** $x=6$ **(b)** $x=\frac{1}{3}$
   (c) $a=5$ **(d)** $x=4,\ x=-4$
5. **(a)** $a=\frac{3}{2}$ **(b)** $x=1$
   (c) $a=4$ **(d)** $x=2$
7. $x=3,\ x=-3$
9. Using either base, the solution is $x=\frac{3}{2}$.

12.2 Graphing Exponential Functions

1. **(a)** exponential **(b)** $(-\infty, +\infty)$
   (c) $(0, +\infty)$ **(d)** $(0, 1)$
   (e) none **(f)** the x-axis, $y=0$
   (g) increasing **(h)** 3
3. **(a)** move 3 left
   (b) move 1 right and 4 up
   (c) move 1 left and 2 down
   (d) move 4 left and 5 down
5. **(a)** $y=\left(\frac{3}{2}\right)^x$ **(b)** $y=\left(\frac{2}{3}\right)^{-x}$
7. **(a)** $(0, 5),\ y=0$
   (b) $\left(0, \frac{4}{3}\right),\ y=0$
   (c) $(0, -3),\ y=0$
9. **(a)** 3.43
   (b) 5,130,000
   (c) 1.77

12.3 Exponential Functions

1. $9,151.43
3. The effective interest rates are 6.6972%, 6.7126%, and 6.7089%, respectively, so Series B is the best investment.
5. In 1990, there were 5,280 subscribers. In 1996, there were 45,150. This growth rate cannot continue indefinitely, because in about 50 years, the number of subscribers would eventually exceed the entire population of the United States.
7. $\frac{32}{243} A_0$
9. The house will be worth $152,222.05.
11. 40 bacteria
13. Graph the function $A = \$23,500(1 - 20\%)^t$.

Examine the graph to find the point where y is approximately equal to $11,750. The corresponding x-value is approximately 3 years.

12.4 Properties of Logarithms

1. *examples:*
   (a) the number of times you have to multiply 10 by itself to get 100 is 2.
   (b) because $\log_b x$ is the exponent to which b is raised to get x
   (c) since the number of times you have to multiply b by itself to get b is 1
5. **(a)** $3 + \log_3 x$ **(b)** $2 - \log x$
   (c) $\frac{1}{2}\log_5 27$ **(d)** $1 + \log a + \log b$
7. **(a)** $\log_2 \frac{x+1}{x}$ **(b)** $\log_3 \frac{x}{8(x+2)}$
   (c) $\log x^2\sqrt{y}$ **(d)** $\log_b \frac{x}{y}$
   (e) $\log_2 \frac{x^3 z^7}{y^5}$ **(f)** $\log_b \frac{\sqrt{x+2}}{y^3 z^7}$
9. **(a)** 2.6609 **(b)** 3
   (c) 2.3218 **(d)** 0.3391
   (e) −0.3391 **(f)** −0.8218
11. *example:* The [log] key on a calculator applies the logarithmic function with a base of 10, so we can use the change of base formula to find $\log_2 7$. The key sequence is 7 [log] [÷] 2 [log] [=].
13. **(a)** In the second line, the multiplication should be addition. The expansion is $\log x + \frac{1}{5}\log y - \log z$.
    (b) In the first line, the change-of-base formula should have been used to obtain $\frac{\log \sqrt{7}}{\log \sqrt{4}}$. The final result is 1.4037.
    (c) 1.0792 should have been subtracted, not added. The answer is −0.7782.

12.5 Logarithmic Equations

1. **(a)** product
   (b) quotient
   (c) power
   (d) base
   (e) exponential
   (f) logarithmic
3. **(a)** log; log 2

(b) $2x - 3$

5. (a) No, because log (−1) does not exist.
   (b) Yes, $w = -2$ is a solution.

7. (a) $x = 10$
   (b) $y = 11$
   (c) $a = 4$
   (d) $b = 625$

9. (a) $a = 1.27024$
   (b) $b = 0$
   (c) $x = 1.80735$
   (d) $y = -8.21440$
   (e) $x = 0, 2.15338$
   (f) $a = \pm 1.07385$

11. The error is in the line $x + 2 \log 3 = x \log 6$.
    The line should be $(x + 2) \log 3 = x \log 6$.
    The correct answer is $x \approx 3.16993$.

12.6 Graphing Logarithmic Functions

1. $y = \log x$ and $y = 10^x$ are inverses of one another.

3. (a) translate 2 left
   (b) reflect over the x-axis
   (c) translate 2 down

5. (a) $(0, +\infty)$
   (b) $(-\infty, +\infty)$
   (c) none
   (d) $(1, 0)$
   (e) increasing
   (f) $y = 1$

7. (a) $Q(b^a, a)$

9. (a) 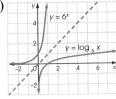 (b)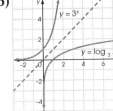

   (c) 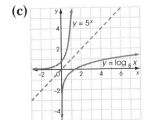 (d)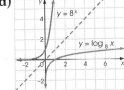

11. (a) increasing
    (b) decreasing
    (c) decreasing
    (d) increasing

12.7 Applications of Logarithms

1. *example:* The logarithm (with base *b*) of a number is equal to the exponent to which the base must be raised in order to get the given number.

3. (a) 32   (b) 40
   (c) 46   (d) 47

5. 50

7. 8 years

9. 1,000

11. 18 years 7 months (223 compounding periods)

12.8 Base-*e* Exponential Functions

3. (a) 54.60
   (b) 0.04
   (c) 20.09
   (d) 1,210.29
   (e) 3,250.76

5.

7. $13,375.68

9. $2,621.40

11. 51 cattle

12.9 Base-*e* Logarithms

3. (a) inverse
   (b) the y-axis
   (c) $(0, \infty)$
   (d) 1
   (e) $e^y = x$
   (f) undefined

5. The inverse of a function f is a function g such that $f(g(x)) = x$. So the statement that $y = \frac{1}{x}$ is an inverse function to itself means that if $y = \frac{1}{x}$ is substituted for x in the equation, then we get $y = x$, the definition of an inverse.

7. 13.9 years

9. (a) $x = e^3$
   (b) $x = \ln(1.5) + 1$

11. $2 \ln x + 3 \ln y - 5 \ln z$

13. $\ln\left(\dfrac{A}{P}\right) = rt$

15. (a) $2,158.92   (b) $2,191.12
    (c) $2,219.64   (d) $2,225.54

Picture Credits

Data in Tables 1: NASA; SSC Programme Ltd.

Data in Tables 2: Tom Dart; First Light © 95 Paul Barton

The Real Number System: CORBIS-BETTMANN; © Imtek Imagineering/Masterfile; *Solving Problems*: Link-Belt/ESS Ltd; Brown Brothers, Sterling, PA

Operations with Real Numbers: CORBIS-BETTMANN; First Light

Rational Exponents: First Folio; Environment Canada

Simplifying Radical Expressions: CORBIS-BETTMANN-UPI; Dick Hemingway

Multiplying and Dividing Radical Expressions: NASA; Particle Data Group, Lawrence Berkeley National Laboratory

Simplifying Algebraic Expressions: CORBIS/Gianni Dagli Orti

Solving Linear Equations and Formulas: Hawai'i Visitors and Convention Bureau/Robert Coello; Mactutor History of Mathematics Archive/University of St. Andrews; Robert Oliver Photography

Applications of Equations: Dick Hemingway; Toronto Sun/Michael Peake

The Rectangular Coordinate System: California Travel and Tourism; NASA

Multiplying Polynomials: Masterfile; CORBIS-BETTMANN

The Greatest Common Factor and Factoring by Grouping: Stock Montage Inc.; CORBIS

Factoring Trinomials and Difference of Squares: Bombardier Inc.; Manitoba Hydro

Sum and Difference of Two Cubes: CORBIS; CORBIS

Dividing Polynomials by Binomials: Lenore Blum/MSRI; CORBIS-BETTMANN

Rational Expressions: Finding Equivalent Forms: Dick Hemingway; Culture Branch, City of Toronto

Non-permissible Values: CORBIS-BETTMANN

Multiplying and Dividing Rational Expressions: CORBIS-BETTMANN

Solving Rational Equations: Dick Hemingway; Ariel Muller Design

Differences Between Relations and Functions: CORBIS-BETTMANN; CORBIS-BETTMANN

Describing Functions: Dairy Farmers of Ontario; CORBIS-BETTMANN

Function Notation: Dick Hemingway; Dick Hemingway;

Domain and Range of Relations: CORBIS-BETTMANN; CORBIS-BETTMANN

Linear Functions: First Light; Donna McLaughlin/First Light

Direct & Partial Variation: Toronto Sun; Paul Tracy/Toronto Sun

Rate of Change and Slope of a Line: Dick Hemingway; Courtesy of Helena Mitasova, University of Illinois at UrbanaChampaign

Solving Linear Systems by Graphing: Masterfile; First Light

Solving Linear Systems by Elimination: Stark Images Photography; Toronto Sun/Paul Henry

Solving Three Equations in Three Variables: Courtesy of Canadian Wheat Growers

Solving Linear Systems by Matrices: Mactutor History of Mathematics Archive/University of St. Andrews; Dick Hemingway

Solving Linear Systems by Determinants: Mactutor History of Mathematics Archive/University of St. Andrews; Dick Hemingway

Solving Systems of Linear Inequalities: Dairy Farmers of Ontario

Linear Programming: Masterfile; Wild Woods Expeditions

Functions and Operations: NASA; NASA

Inverse Functions: Jackie Lacoursiere; Aracor

Graphs of Quadratic Functions: Marquis by Waterford; Courtesy of Tedmonds Satellite & Cellular

Completing the Square: Mike Addie, Riverside CA; Six Flags, California; First Light

Translations of Quadratic Functions: Adam G. Sylvester/Photo Researchers; David Hamilton

Solving Quadratics by Graphing: Raleigh; Masterfile

Solving Quadratics by Factoring: BC Tourism; CORBIS-BETTMANN

Complex Numbers: Mactutor History of Mathematics Archive/University of St.

Andrews

The Quadratic Formula: Dick Hemingway; African Photo Safari

Using Discriminants and Graphs: NASA

Graphs of Polynomial Functions: Dangerless Aerial Sports Club; M.I.T

Graphs of Rational Functions: NASA; Dick Hemingway

Absolute Value Equations: Jackie Lacoursiere; Jackie Lacoursiere

Radical Equations: CORBIS/Lawrence Manning; CORBIS/Comnet Limited; Telesat Canada Satellite; Mactutor History of Mathematics Archive/University of St. Andrews

Rational Equations: Langford Canoe; First Light

Rational Inequalities 1: Canadian Home Workshop Magazine; CORBIS/Historical Picture Archive

Rational Inequalities 2: Reuters/Joe Traver/Archive Photos; Lord Egremont, Petworth House Archives

Complex Fractions: Mactutor History of Mathematics Archive/University of St. Andrews

Exponential Equations: CORBIS/Ray Bird, Frank Layne Picture Agency; CORBIS/Layne Kennedy

Graphing Exponential Functions: Intel Corporation; Dick Hemingway

Exponential Functions: Toronto Sun/Michael Peake; CORBIS-BETTMANN

Properties of Logarithms: Mactutor History of Mathematics Archive/University of St. Andrews; Dick Hemingway

Logarithmic Equations: California Institute of Technology; Reuters/Claro Cortes/Archive Photos

Graphing Logarithmic Functions: Mactutor History of Mathematics Archive/University of St. Andrews; Mactutor History of Mathematics Archive/University of St. Andrews

Applications of Logarithms: Archive Photos; Reuters/CORBIS-BETTMANN

Base-e Exponential Functions: Dick Hemingway; Freefall Adventures, Florida School of Accelerated Freefall

Base-e Logarithms: Mactutor History of Mathematics Archive/University of St. Andrews; CORBIS

Index